Development and Local Knowledge

There is a change happening in the practice of applied anthropology. A new field of 'indigenous knowledge' is emerging which aims to make local voices heard and ensure that development initiatives meet the needs of ordinary people. Indigenous knowledge, an aspect of participatory approaches to development, offers an alternative to schemes and strategies that are imposed on lesser developed countries by international agencies and state organizations.

The philosophy behind the indigenous knowledge initiative is straightforward. It is based on the belief that effective assistance will benefit from some understanding of local knowledge and practices, by promoting culturally appropriate and sustainable interventions.

Achieving the aims of indigenous knowledge presents significant challenges. *Development and Local Knowledge* focuses on two major issues that might be addressed by anthropology: first, the proper definition of indigenous knowledge – what is it, who should define it and what are the implications, particularly political, of any definition? second, the advancement of methodologies appropriate to the exploitation of indigenous knowledge – how can development access it sympathetically, who should control its use and how? While accepting that working with local knowledge is never easy, the authors offer ways of advancing the relationship between local knowledge and development, and of furthering anthropology's key role in development processes.

Alan Bicker is Lecturer and Research Fellow at the University of Kent. **Paul Sillitoe** is Professor of Anthropology at the University of Durham. **Johan Pottier** is Professor of African Anthropology at the School of Oriental and African Studies, London. Together they are the editors of *Participating in Development* (2002) and *Negotiating Local Knowledge* (2003).

Studies in environmental anthropology
Edited by Roy Ellen
University of Kent at Canterbury, UK

This series is a vehicle for publishing up-to-date monographs on particular issues in particular places which are sensitive to both sociocultural and ecological factors. Emphasis will be placed on the perception of the environment, indigenous knowledge and the ethnography of environmental issues. While basically anthropological, the series will consider works from authors working in adjacent fields.

Volume 1 A Place Against Time
Land and environment in the Papua New Guinea highlands
Paul Sillitoe

Volume 2 People, Land and Water in the Arab Middle East
Environments and landscapes in the Bilâd as-Shâm
William Lancaster and Fidelity Lancaster

Volume 3 Protecting the Arctic
Indigenous peoples and cultural survival
Mark Nutall

Volume 4 Transforming the Indonesian Uplands
Marginality, power and production
Edited by Tania Murray Li

Volume 5 Indigenous Environmental Knowledge and its Transformations
Critical anthropological perspectives
Edited by Roy Ellen, Peter Parkes and Alan Bicker

Volume 6 Kayapó Ethnoecology and Culture
Darrell A. Posey, edited by Kristina Plenderleith

Volume 7 Managing Animals in New Guinea
Preying the game in the highlands
Paul Sillitoe

Volume 8 Nage Birds
Classification and symbolism among an Eastern Indonesian people
Gregory Forth

Volume 9 Development and Local Knowledge
New approaches to issues in natural resources management, conservation and agriculture
Alan Bicker, Paul Sillitoe and Johan Pottier

Development and Local Knowledge

New approaches to issues in natural resources management, conservation and agriculture

Edited by Alan Bicker, Paul Sillitoe and Johan Pottier

LONDON AND NEW YORK

First published 2004
by Routledge
2 Park Square, Milton Park, Abingdon, Oxon, OX14 4RN

Simultaneously published in the USA and Canada
by Routledge
270 Madison Ave, New York NY 10016

Routledge is an imprint of the Taylor & Francis Group

Transferred to Digital Printing 2006

Typeset in Galliard by Wearset Ltd, Boldon, Tyne and Wear

British Library Cataloguing in Publication Data
A catalogue record for this book is available from the British Library

Library of Congress Cataloging in Publication Data
A catalog record for this book has been requested

ISBN 0–415–31826–2

Printed and bound by CPI Antony Rowe, Eastbourne

Contents

Illustrations

Figures

Tables

Contributors

Mahbub Alam, Bangladesh Resources Centre for Indigenous Knowledge (BARCIK), 3/7 Block-D, Lalmatia, Dhaka – 1207, Bangladesh.

Julian Barr, Natural Resources, Environment and Social Development Work Group, Information, Training and Development (ITAD) Ltd, Lion House, Ditchling Common, Ditchling, Hassocks, West Sussex BN6 8SG, UK.

Alan Bicker, Anthropology Department, University of Kent, Eliot College, Canterbury CT2 7NS, UK.

Greg Cameron, Department of Political Science, University of Asmara, Box 1220, Asmara, Eritrea.

Colin Filer, Resource Management in Asia-Pacific Program, RSPAS, ANU, Canberra ACT 0200, Australia.

Michael Fischer, Anthropology Department, University of Kent, Eliot College, Canterbury CT2 7NS, UK.

Terence Hay-Edie, Laboratoire d'Anthropologie Sociale, Collège de France, Paris, France.

Ilse Köhler-Rollefson, League for Pastoral Peoples, Pragelatostr. 20, 64372 Ober-Ramstadt, Germany.

Constance McCorkle, CMC Consulting, 7767 Trevino Lane, Falls Church, VA 22043, USA.

Johan Pottier, Anthropology Department, School of Oriental and African Studies, Thornhaugh Street, Russell Square, London WC1H 0XG, UK.

Hans Siebers, Utrecht School of Governance, Utrecht University, Bijlhouwerstraat 6, 3511 ZC Utrecht, The Netherlands.

Paul Sillitoe, Anthropology Department, University of Durham, 43 Old Elvet, Durham DH1 3HN, UK.

Paul Spencer, International African Institute, SOAS, Thornhaugh Street, Russell Square, London WC1H 0XG, UK.

Pamela J. Stewart, Department of Anthropology, University of Pittsburgh, 3H01 Posvar Hall, Pittsburgh, PA 15260, USA.

Veronica Strang, School of Social Sciences, Auckland University of Technology, Private Bag 92006, Auckland 1020, New Zealand.

Andrew Strathern, Department of Anthropology, University of Pittsburgh, 3H01 Posvar Hall, Pittsburgh, PA 15260, USA.

Acknowledgements

This volume was spawned by the Association of Social Anthropologists' conference on Indigenous Knowledge and Development at the School of Oriental and African Studies, University of London, in 2000. We therefore thank its contributors for their commitment to the issues raised, and Routledge for their support throughout.

Chapter 1

Introduction

Hunting for theory, gathering ideology

Paul Sillitoe and Alan Bicker

Development agencies are increasingly sympathetic to the proposition that indigenous knowledge should feature in the planning and implementation of programmes. They are ever more receptive with the failure of many development projects, the emergence of 'stakeholder participation' and the advancement of policies targeting the poor who depend heavily on indigenous strategies. Such work, briefly, seeks to facilitate a larger role for local peoples' knowledge and aspirations in interventions planned for their regions.

There is some academic debate over the propriety of the term indigenous knowledge, and by extension the correctness of engaging in such work (Agrawal 1995; Antweiler 1998; Ellen and Harris 2000), an argument taken up by Fisher in this book. Some are unhappy at the use of the word indigenous, on the grounds that it is difficult to determine the status of indigene (e.g. Colchester 2002; McIntosh 2002), and suggest other terms such as local or traditional knowledge. Whatever term we use, there are objections. It is no easier to define local or traditional than indigenous. And indigenous is the label, for better or for worse, that has caught on in development circles. We have taken up the debate over the definition of the term indigenous knowledge, and gone even further to question the meaning of the term development, in a companion volume to this one, entitled *Participating in Development: Approaches to indigenous knowledge* (Sillitoe, Bicker and Pottier 2002), which originated from the same conference. This was the millennial Association of Social Anthropologists of the United Kingdom and Commonwealth (ASA) Conference (at London University's School of Oriental and African Studies, April 2000), which aimed to further debate about the place of indigenous knowledge in development.

We should not claim to have laid the definitional debate to rest and the term indigenous knowledge remains contentious. Indeed, it is a recurrent theme throughout the contributions to this volume. Veronica Strang (Chapter 6) points out that definitions of indigenous knowledge that neatly fit Western scientific models necessarily confound the very essence of emic definitions of that knowledge. Colin Filer's observation (Chapter 5) that most definitions of indigeneity emphasize the subordinate political status of the peoples concerned, means that their indigenous knowledge is itself demeaned. Which brings us to Strathern and Stewart's enquiry in Chapter 4: what do we mean by indigenous? And how do we distinguish knowledge from performance? Nonetheless, we must give some indication of what it encompasses. indigenous knowledge, to define it concisely, is any understanding rooted in local culture. It includes all knowledge held more or less collectively by a population that informs interpretation of things. It varies between societies. It comes from a range of sources, is a dynamic mix of past tradition and present innovation with a view to the future.

Indigenous knowledge challenges

The incorporation of indigenous knowledge in development presents us with a number of interesting challenges beyond deciding the content of the field and arguing over the rightness of engaging in such work. It is not straightforward work. It is necessary to proceed cautiously, aware of the difficulties. They demand attention to integrate indigenous knowledge into the development process. We need to formulate strategies that meet the demands of development – cost-effective, time-effective, generating relevant insights, readily intelligible to non-experts, etc. – while not downplaying the difficulties so as to render the work effectively valueless. While development project managers will assess attempts to advance on current techniques according to their resource effectiveness, they should set these demands against the range of information collected and its reliability.

At first sight indigenous knowledge work seems straightforward enough, we just have to ask some local culture-bearers what they think. But we soon run into cross-cultural problems that challenge what we think we know, as any anthropologist will confirm. The task of sympathetically accessing concepts in local usage, and conveying something about them, is large. Knowledge is diffuse and communicated piecemeal in everyday life. As Hans Siebers (Chapter 3) shows in his discussion of the management or what he terms the 'creolization' of indigenous knowledge, there is often no consensus among the natives, and local stakeholder knowledge is not homogenous. People transfer much knowledge through practical experience and are unfamiliar with expressing all that they know in words. They may also carry knowledge, and transfer it between generations, using alien idioms featuring symbols, myths, rites and so on. Translating what we hear into foreign words and concepts further misconstrues whatever it is that we manage to comprehend about others' views and actions. Understanding is inevitably limited given our outsider perspective, development-oriented indigenous knowledge work is no different from any other ethnographic enquiry in this respect (see another companion volume to this one, Pottier, Bicker and Sillitoe (2003) for a full discussion of the negotiated nature of indigenous knowledge, and the sometimes conscious translations to which indigenous knowledge-as-we-know-it can be subjected).

The advancement of interdisciplinary work is central to indigenous knowledge research, particularly when combining the technical know-how of natural scientists with the empathy of social scientists. An integrated perspective implies a willingness to learn from one another, as well as from local people. The indigenous knowledge component of any research and/or development project should not necessarily dominate. There must be a genuine two-way flow of ideas and information between all parties. Motivation depends in considerable measure on fostering consensus decisions, joint ownership and open debate: issues eloquently addressed by Michael Fischer (Chapter 2) and Paul Spencer (Chapter 11). Indeed, Spencer's discussion of Chamus negotiating skills and practices in particular, and those of East African pastoralists in general, provides an object lesson for many natural and social scientists (including anthropologists). In addition to exploring the meanings attributed to indigenous knowledge and development, the above *Participating in Development* volume addresses these interdisciplinary issues. indigenous knowledge research should maintain a wide socio-cultural perspective to contextualize the narrowly focused work of technical specialists. In science-speak, we cannot understand cultures by looking at individual parts in isolation, as complex systems they manifest emergent properties that we can only see when all the parts are working together. It is not possible to predict which cultural domains might relate intimately with others, often unexpected practices impinge on one another. Michael Fischer (Chapter 2) illustrates the need for caution here in his insightful description of differing perceptions of what constitutes a useful potato – in his Pakistani example, it was not just the Swiss advisers who differed from their Pathan clients but the Pakistani government's own experts.

A key problem is facilitating meaningful communication between development personnel and local people to establish what each has to offer, informing science with ethnographic findings about people's knowledge and locals about the scope of science and what it might offer, so that they can better understand the alternatives available in addressing problems, so realizing the comparative advantages of each. As Ilse Kôhler-Rollefson and Constance McCorkle (Chapter 9) show in their plea for domestic animal diversity. This is not to imply that science is better, for as the café owner in Zadie Smith's novel *White Teeth* comments, 'Science ain't no different from nuffink else, is it? I mean, when you get down to it. At the end of the day, it's got to please the people, you know what I mean?' It is however the system of knowledge which underpins the technological capacity that informs development. The promotion of more effective participation in the identification and researching of problems can only be achieved so far as awareness, knowledge and socio-political barriers will allow. We have to seek ways forward that allow both outsiders and insiders to contribute as necessary, as several of the chapters to this book argue, balancing between technocrats defining the problem/constraint, which can be arrogant and ethnocentric, and the local people doing so, which hits cultural barriers that thwart scientific research. The objective is equitable negotiation, a central tenet of participatory development. The negotiations may be complex but development initiatives are more likely to be locally appropriate and sustainable. The presentation of indigenous knowledge in a manner accessible to others, such that they can see its relevance to their work, means avoiding jargon-loaded and obscure accounts, while not overlooking insights gained in cross-cultural research, often using subtle arguments. There is a need to avoid oversimplification of complex issues, inviting distortion and misrepresentation in the search for user-friendly accounts.

It is necessary to promote a collaborative atmosphere in which neither scientific nor local interests feel threatened, assuring all parties that they have a role in negotiations, with vital skills and knowledge. This implies demonstrating how awareness of indigenous knowledge will improve the relevance of development work and vice versa. The second volume resulting from the ASA millennial conference, entitled *Negotiating Local Knowledge: Identity, power and situated practice in development intervention* (Pottier, Bicker and Sillitoe, 2003), explores further the issues surrounding communication and negotiation. Negotiation also goes on *within* local communities: people constantly reappraise what they know and modify it in the light of experience, including political interests. The dynamism of indigenous knowledge exacerbates the difficulties we experience in representing it. As indigenous knowledge is neither static nor uniform, it cannot be documented once-and-for-all, but is subject to continual negotiation between stakeholders. As Greg Cameron shows in Chapter 8, the dynamics that constitute indigenous knowledge are often subject to murky power struggles within and between the state and NGOs, frequently to the detriment of local perceptions and aspirations. If we hope to accommodate to the dynamic nature of local knowledge we need an iterative research strategy, closely linking development interventions to on-going indigenous knowledge investigations. After all, development aims to accelerate change, dramatically modifying indigenous knowledge with scientific perspectives.

The time scale required in ethnographic research is normally considerable which presents problems in development contexts with short-term politically driven considerations demanding quick returns. Managers need to understand that indigenous knowledge research is usually long term, as the research that informs the contributions to this book show, some of the authors have decades of experience working in the regions they describe. It can take several years, not months or weeks, for someone unacquainted with a region to achieve meaningful insight into local knowledge and practices, and from this perspective to inform development projects. The understanding that can be accomplished in a single project cycle will be of a different order. While some indigenous knowledge research may be attempted in short time frames, it is necessary to be aware of the costs of necessary compromises. It is not just a question of the time it takes to learn language, cultural repertoire, social scenario and so on, but also the investment needed to win the trust and confidence of people who frequently have reason to be extremely suspicious of foreigners and their intentions.

The one-off nature of indigenous knowledge research also hampers its deployment in development, impeding the formulation of generalizations that might inform wider policy and practice. Its small-scale, culturally specific and geographically local nature hinders the advancement of an integrated approach. This is possibly a red herring given the variety of knowledge traditions worldwide, their internal variations regarding individuals' assorted understandings and their constant revision over time. This variation makes generalization potentially dangerous, imputing ideas elsewhere that may be inappropriate. Nonetheless it is argued that we need to evolve principles that will facilitate a degree of reliable generalization from indigenous knowledge research, to go beyond local case studies that are not cost effective to replicate in large numbers. It is to this issue that the contributions in this volume relate, for they are in the case history tradition.

Indigenous knowledge theory?

Indigenous knowledge research as currently conducted is largely ethnographic documentation of others' environmental relations and livelihood systems. It is not analytical regarding these systems nor is it framed to identify and help address scientifically researchable constraints that limit their productivity. It has proved effective in some small-scale NGO work conducted by those working close to a few communities, notably featuring limited appropriate technology interventions, but has so far had little large-scale impact, failing to inform wider understanding of problems, and regional policy and practice. Even in NGO contexts there is scope for a deeper anthropological awareness among those who advocate both indigenous knowledge and participatory approaches, but in the context of bilaterally and multilaterally funded research and development (e.g. DFID, FAO, USAID, etc.) there is an urgent need for it, or else there is the danger of indigenous knowledge advocacy appearing amateurish, produced by social scientists ignorant of technical research. We need a professional edge to penetrate the scientific research establishment. The absence of a coherent indigenous knowledge intellectual framework that might interface effectively with science and technology contributes to natural scientists failing to appreciate it and see how it might inform their research. Consequently indigenous knowledge research appears to contribute to the accumulation of exotic ethnographic documentation and databases that are sterile and undynamic from a developmental perspective, even potentially disempowering people by representing their knowledge in ways inaccessible to them and beyond their control, and maybe even infringing their intellectual property rights. As Greg Cameron shows (Chapter 8) there are various levels at which this disempowering process can take place locally. NGOs may accept the arguments of international indigenous rights advocates, which encompass indigenous knowledge, to the detriment of local resistance to top-down development policies.

These shortcomings of indigenous knowledge research suggest that we need to formulate some theory to give it coherence and structure. This is what development agencies currently seek, a general approach that they can commission globally, like crop breeding or soil fertility research, to help solve problems. But after giving it some thought we are not convinced that this is the way forward. It may even be a wrong-headed endeavour, not only because of the diversity of knowledge traditions globally and the danger of ethnocentrically imposing some uniform theoretical straitjacket on them, but also the incomparability of scientific theory with whatever might masquerade as an indigenous knowledge theory (even if, as Michael Fischer proposes in Chapter 2, applications of scientific knowledge and indigenous knowledge may undergo processes with similar qualities. Anthropology long ago gave up trying to formulate a theory for humankind analogous to the powerful theory of natural science, following efforts such as those of functionalism, which hypothesized that socio-cultural institutions exist to ensure human survival – in initial formulations ensuring biological needs are met and subsequent reformulations focusing on the maintenance of orderly social environments. But this was less a theory in the scientific sense than an analogy that drew on anatomical theory, likening social customs to biological organs, functioning to keep any society in existence. It came up with nothing like a zoological theory-equivalent to explain the social analogues of heart, lungs, circulation, etc., nor even a way of identifying the counterparts of these organs.

A review of anthropology's attempts at universal cross-cultural generalizations should further warn us against trying to come up with a globally relevant indigenous knowledge theory to match against scientific theory. These are, by and large, grandiose statements of the obvious, often camouflaged in obscure clever-sounding arguments. The early functionalist theory of Malinowski (1944) is a good example, and has the merit of being written in comprehensible prose, which perhaps reveals its naïveté so clearly. It boils down to arguing that cultural arrangements ensure that human beings reproduce, secure food, have shelter, etc. We can embellish our theories with complex exegeses and classifications of various marriage institutions, behavioural taboos, livelihood regimes or whatever – e.g. humans think in binary opposites, communicate in various ways, seek to dominate and exploit others – but these scarcely advance our understanding of humanity in the way that atomic theory affords insights into matter. It seems that we have two meanings of the word theory.

If the idea of anthropological, and by extension indigenous knowledge theory, is incommensurate with the idea of scientific theory, this prompts us to ask 'what is anthropological theory?' While the word theory features prominently in anthropological discourse (to achieve greatness, invent one), we confess that the idea of anthropological theory has long puzzled us. It is not merely the obscure jargon that some writers use to express convoluted ideas (ideas that are often surprisingly straightforward when one translates the jargon), but the notion of theory itself. We use the term freely but do we agree what we mean by it? It is currently popular in the social sciences to deconstruct terms (debate their origin and meaning), such as the word indigenous, and argue over the propriety of their use. Such an exercise in deconstruction of the term theory, as used in anthropology, is perhaps overdue.

If there is no such thing as anthropological theory (or social theory generally) that is analogous to scientific theory, what is there? In science, a theory is a system of understanding that has explanatory and predictive power applied to the inanimate world that manifests consistent regularities. This definition of theory accords with that found in dictionaries, as a 'system of ideas intended to explain something, especially one based on general principles independent of the thing to be explained' (*Oxford English Dictionary*). In social science usage, theory is hotly contested and subject to rapid intellectual fashion changes, it has limited explanatory and virtually no predictive powers. Indeed there is no agreed theory but an endless series of competing theories with their own vocabularies, etc. There is no close definition of terms and no rigorous use of agreed rules by which to manipulate concepts, as we find in the mathematical logic that informs much use of scientific theory. These theories represent attempts to understand very complex phenomena that exist and manifest themselves in human behaviour in its endless variety. They are less theory to assist explanation than ideology that informs interpretation. They express 'currents of opinion' (Kuper 1999). They depend heavily on individuals' values and views of humanity which are subject to never-ending debate and irreconcilable dispute (for example, some people believe that some persons are superior to others and should occupy privileged social positions, whereas other people believe that all humans are equal and none should occupy stations above others – such contradictory positions are what characterizes the human condition, as some Eastern philosophies make clear in their mind games). When taking up a theoretical position, social scientists are more often taking up an ideological position, which will usually associate closely with contemporary concerns in their own culture and social lives. In an anthropological context this can be tantamount to ethnocentricism, imposing one's own culturally informed ideas and beliefs on others. We frown on it. In indigenous knowledge research, as in anthropology generally, we should struggle to avoid it.

Where are we at? If social science ideologies are different from scientific theory, and their use further threatens ethnocentric distortion of others' ideas, beliefs and behaviours, how are we to present indigenous knowledge in a way that it can interface with scientific theory? There is, in short, no single indigenous knowledge theory that might correlate with scientific theory. It is not a universal issue but a local one as chapters in this volume make abundantly clear, especially Fischer (Chapter 2), and Siebers (Chapter 3) (see also Cleveland and Soleri 2002 for an example considering peoples' ideas about crop breeding compared with those of scientific crop breeders). We are concerned with a series of attempts by people in different parts of the world to relate what they know to the scientific understanding that informs development. It is for local people themselves to formulate, to come up with their own theory equivalents to interface with science. This shifts the burden in indigenous knowledge research away from seeking the holy grail of a grand theory, to coming up with methodologies to facilitate people interfacing with science and technology. Almost all of the contributors to this volume represent this aim, but see particularly Sillitoe, Barr and Alam (Chapter 10) in their quest for soil maps that are more relevant to local people. This accords with current postmodern criticism that we should not assume to represent others, an inevitably ethnocentric exercise. The case history approach of the chapters in this book, particularly those of Strang (Chapter 6), Köhler-Rollefson and McCorkle (Chapter 9), and Sillitoe, Barr and Alam (Chapter 10), exemplify this approach to indigenous knowledge work, that each socio-cultural context has to represent its own relationship with scientific theory. The result will be indigenous knowledge research replicated differently over and over at different places and at different times, and as Fischer (Chapter 2) proposes, 'detailed contextualized and finely textured studies, systematic fieldwork, development of theory, applied work and the development of new qualitative and quantitative approaches to study these issues'. This complies with current calls for mid-range theories in the social sciences, away from grand theories of humankind.

Indigenous knowledge ideology

If we are arguing that there is no socio-cultural theory equivalent to scientific theory but rather a range of ideologies, where does this put the search for some general perspective on indigenous knowledge research? It shifts enquiries from a search for an indigenous knowledge theory equivalent to scientific theory, to an ideological debate. This has indeed been going on. The ideological debate is seeking to establish a place for indigenous knowledge in development. It is trying to work out how to get the local voice heard, both by managers and scientists working on projects at the community level, and by politicians and policy makers in agencies at the international level. The broad issue is one of politics. The political dimension is prominent at all levels where we challenge and seek to convince authorities that there are benefits to be gained by giving more opportunity to local communities to determine their own destinies in the light of their knowledge and values. While some of us have been working at the local level to further this, others have been campaigning at the international level.[1] Albeit, as Hans Siebers (Chapter 3) suggests, actual Western policies are increasingly marked by an 'instrumental rationality that does not fit with a plausible plea for such broad objectives'.

Nevertheless, one cannot imagine a more ideological statement than the 1974 Cocoyoc Declaration of the United Nations Environment Programme (UNEP) and Conference on Trade and Development (UNCTAD), which points out that an affluent minority consuming the larger part of the earth's resources while a poor majority struggle to survive, compromises any hope of sustainable development. Two years previously, at Stockholm in 1972, the UN Conference on the Human Environment linked human rights and well-being to environmental sustainability, a connection subsequently recognized by many countries, including those in the Organization for Economic Co-operation and Development (OECD). It lead to many multilateral environmental agreements, such as the Conventions on World Cultural and Natural Heritage (WHC 1972) and International Trade in Endangered Fauna and Flora (CITES 1973), all of which relate to the rights and responsibilities of local people as contributors to sustainable development. An increasing gap between rich and poor together with burgeoning global environmental problems, such as ozone depletion and global warming, led in the 1980s to the linking together of sustainability and development. The United Nations established in 1983 the World Commission on Environment and Development (WCED – the Brundtland Commission) to enquire into environmental issues in developing countries. It involved indigenous people in its work and its report *Our Common Future* (1987) stressed the value of traditional knowledge, called for the empowerment of local communities and protection of their land and resource rights.

The international community further acknowledged that indigenous people should play a key part in sustainable development at the United Nations Conference on Environment and Development (UNCED), in 1992 at Rio de Janeiro, the so-called Earth Summit (Noejovich 2001). Although the intergovernmental discussions formally excluded minority groups and NGOs, they lobbied delegates vigorously and organized a parallel Earth Parliament at Kari-Oca on the outskirts of Rio. They issued the *Kari-Oca Declaration* or the Indigenous

Kari-Oca Declaration
Brazil, 30 May 1992

We, the Indigenous Peoples, walk to the future in the footprints of our ancestors.

From the smallest to the largest living being, from the four directions, from the air, the land and the mountains, the creator has placed us, the Indigenous Peoples upon our Mother the earth.

The footprints of our ancestors are permanently etched upon the lands of our peoples.

We, the Indigenous Peoples, maintain our inherent rights to self-determination. We have always had the right to decide our own forms of government, to use our own laws, to raise and educate our children, to our own cultural identity without interference.

We continue to maintain our rights as peoples despite centuries of deprivation, assimilation and genocide.

We maintain our inalienable rights to our lands and territories, to all our resources – above and below – and to our waters. We assert our ongoing responsibility to pass these on to the future generations.

We cannot be removed from our lands. We, the Indigenous Peoples, are connected by the circle of life to our lands and environments.

We, the Indigenous Peoples, walk to the future in the footprints of our ancestors.

Peoples' Earth Charter on development and the environment, which might stand as a proclamation for indigenous knowledge work-in-development. It demands recognition of self-determination, self-development, rights to land and resources and respect for cultural heritage as necessary to sustainable development. Rio furthered indigenous peoples' environmental rights considerably, adopting a number of legal instruments to protect rights to traditional environmental knowledge and management and conservation practices. The Rio Declaration (Principle 22) notes the part indigenous people have to play in the drive for sustainable development: 'Indigenous people and their communities, and other local communities, have a vital role in environmental management and development because of their knowledge and traditional practices. States should recognise and duly support their identity, culture and interests and enable their effective participation in the achievement of sustainable development.' The intergovernmental sustainable development plan of action 'Agenda 21' addresses indigenous communities in Chapter 26 calling for nations to adopt appropriate policies, recognize indigenous values, knowledge and resource management practices, and promote local participation in sustainable development strategies.

Subsequently, the Convention on Biological Diversity came into force in 1994, Articles 8j, 10c and 10d recognizing indigenous land rights, environmental knowledge and informed consent in relation to its use, right to manage lands and resources according to customary law, and equitable sharing of benefits from resource use. In 1995 the Intergovernmental Panel on Forests was established to further action on forests and indigenous representatives participated in sessions and pressed for participatory forest policies, and in 2000 the United Forum on Forests came into being with a mandate to serve as an open global forum on forest policy (Griffiths 2001). In this volume, Filer presents an entertaining discussion of the implications of forest conservation locally, in Papua New Guinea. Some international indigenous peoples' bodies have established themselves to take these initiatives forward, such as the International Indigenous Forum on Biodiversity which the Conference of Parties working on the implementation of the Convention on Biological Diversity recognizes as an advisory body, and the International Indigenous Forum on Climate Change which seeks to lobby the Framework Convention on Climate Change (1994) that aims to stabilize greenhouse gas emissions at levels that will not interfere with the world's climate.

Action since Rio has been disappointing regardless of conventions and declarations of principle recognizing that indigenous peoples should have a role in sustainable development. Translating words into action, locally through to internationally, remains an enormous challenge (Greenpeace International 2002). Governments lack the political will to sign up, and agreements, few if any of which have been honoured adequately, reaffirm the sovereignty of nation states over natural resources – often at the expense of minority populations (Lâm 2000) – to be traded as economic assets on commodity markets. The Johannesburg World Summit on Sustainable Development in 2002 was seen as an opportunity to turn rhetoric into reality, although the Indigenous Peoples' Caucus, commenting on the Summit's Draft Plan of Implementation, talked about it taking 'a few steps back' from the Rio commitments. It concludes that its participation has not resulted in 'substantive commitments' that might improve the 'well-being of Indigenous Peoples', noting bitterly that 'the negotiated agreements by governments, promises our peoples more of the same mining, energy-production and privatization of water and social services, this time labelled as "poverty eradication" and "sustainable development".'

It was agreed that the World Summit should not aim to produce a new set of policies and conventions, but should rather review the implementation of the Rio instruments. The recognition of indigenous peoples as an Agenda 21 'major group' afforded them a role in Summit preparation, to comment officially on the implementation of sustainable development commitments since Rio and priorities for the future. They have expressed their worries (in the Indigenous Peoples' Caucus Statement for the Multi-Stakeholder Dialogue on Governance, Partnerships and Capacity-Building) that the Summit will continue contemporary development policies with all their inequalities. At the Fourth Summit Preparatory Committee meeting on Bali, indigenous participants issued the Indigenous Peoples' Political Declaration, which confirms their disappointment and emphasizes the steps they think necessary for sustainable development.

> The results of the negotiations to gain consensus and mutual support for the protection of the environment and sustainable development have been very discouraging. We are disappointed that our fundamental rights and the specific language of INDIGENOUS PEOPLES have not been honoured. We fear that our territories and the natural world will continue to be plundered by governments and corporations... For as long as you continue to make war against Mother Earth there can never be peace. Humanity must work together, not just for survival, but for quality of life based on ethical, cultural and spiritual values to protect the sacred inter-relatedness of life that serves us all. We remind us all of our responsibilities to future generations... We reaffirm that self-determination and sustainable development are two sides of the same coin.
>
> (Tebtebba Foundation 2002)

The International Indigenous Peoples' Summit on Sustainable Development held at Kimberley in South Africa in August 2002, to coincide with the Johannesburg World Summit, formally reaffirmed the *Kari-Oca Declaration* and expressed concern at the lack of progress made during the previous decade to implement the Rio agreements. Nonetheless paragraph 25 of the Johannesburg Summit Political Declaration importantly states 'We reaffirm the vital role of the indigenous peoples in sustainable development'. *The Kimberley Declaration* puts their case eloquently.

The Kimberley Declaration
South Africa, 20–23 August 2002

Today we reaffirm our relationship to Mother Earth and our responsibility to coming generations to uphold peace, equity and justice.

As peoples, we reaffirm our rights to self-determination and to own, control and manage our ancestral lands and territories, waters and other resources. Our lands and territories are at the core of our existence – we are the land and the land is us; we have a distinct spiritual and material relationship with our lands and territories and they are inextricably linked to our survival and to the preservation and further development of our knowledge systems and cultures, conservation and sustainable use of biodiversity and ecosystem management.

We have the right to determine and establish priorities and strategies for our self-development and for the use of our lands, territories and other resources. We demand that free, prior and informed consent must be the principle of approving or rejecting any project or activity affecting our lands, territories and other resources.

We are the original peoples tied to the land by our umbilical cords and the dust of our ancestors. Our special places are sacred and demand the highest respect. Disturbing the remains of our families and elders is desecration of the greatest magnitude and constitutes a grave violation of our human rights.

The national, regional and international acceptance and recognition of Indigenous Peoples is central to the achievement of human and environmental sustainability. Our traditional knowledge systems must be respected, promoted and protected; our collective intellectual property rights must be guaranteed and ensured. Our traditional knowledge is not in the public domain; it is collective, cultural and intellectual property protected under our customary law. Unauthorized use and misappropriation of traditional knowledge is theft.

Economic globalization constitutes one of the main obstacles for the recognition of the rights of Indigenous Peoples. Transnational corporations and industrialized countries impose their global agenda on the negotiations and agreements of the United Nations system, the World Bank, the International Monetary Fund, the World Trade Organization and other bodies which reduce the rights enshrined in national constitutions and in international conventions and agreements. Unsustainable extraction, harvesting, production and consumption patterns lead to climate change, widespread pollution and environmental destruction, evicting us from our lands and creating immense levels of poverty and disease.

We are determined to ensure the equal participation of all Indigenous Peoples throughout the world in all aspects of planning for a sustainable future with the inclusion of women, men, elders and youth. Equal access to resources is required to achieve this participation.

We urge the United Nations to promote respect for the recognition, observance and enforcement of treaties, agreements and other constructive arrangements concluded between Indigenous Peoples and States, or their successors, according to their original spirit and intent, and to have States honor and respect such treaties, agreements and other constructive arrangements.

The international bodies representing the interests of indigenous populations define these more narrowly than we do when we talk about indigenous knowledge in development contexts, which returns us again to issues of definition. They define indigenous in Article 1 of the International Labour Organization's (1989) Convention 169, which regards people 'as indigenous on account of their descent from the populations which inhabited the country, or a geographical region to which the country belongs, at the time of conquest or colonization or the establishment of present state boundaries and who, irrespective of their legal status, retain some or all of their own social, economic, cultural and political institutions.' In development contexts, on the other hand, we think of any community that relies on local resources and ways as indigenous, regardless of its historical status. Nonetheless, any advances made by these indigenous interest groups in international forums regarding recognition of such populations' rights are likely to benefit all poor rural communities, including migrants, in gaining a more meaningful voice in the planning and implementation of development initiatives in their regions, as is amply illustrated by Terence Hay-Edie's discussion of the role of UNESCO (Chapter 7). There are no other bodies currently fighting for their interests, other than ideologically motivated, but politically weak, academics engaging in action research.

The dominance of international economic and financial bodies (such as the World Trade Organization, World Bank, International Monetary Fund and the World Intellectual Property Organization that are facilitating so-called economic globalization) are undermining the Rio agreements on sustainable development and effectively sidelining indigenous peoples. As the Indigenous Peoples' Caucus Statement notes, development should be 'about addressing social and power relationships, and about how these relationships impact on our relations with the Earth. The contemporary world is characterised by deep imbalances in our social relations, of gross inequalities between nations and within societies, manifested by huge disparities in consumption... Governance structures for sustainable development must strive for greater democratisation, transparency, equity, and accountability in order to achieve better outcomes.'

At a time when the unchecked commercial exploitation of the environment in the name of naked profit continues recklessly to damage our global commons – with atmospheric pollution destabilizing the world's climate, water pollution and despoliation of the oceans, industrial agricultural production poisoning the land and increasingly threatening health with dubious food products, reckless practices depleting soil resources, etc. – the wisdom of indigenous peoples' pronouncements is evident. It is an iniquitous system condemning millions to lead poverty stricken lives. These are real, not theoretical problems. There cannot be a more ideological call to action.

As can be seen from the summaries below, the contributions to this volume demonstrate the support that the indigenous knowledge in development initiative gives to the demands of the international forums representing indigenous peoples for the recognition of their right to self-determination and representation through their own institutions (Lâm 2000). These include strengthening local peoples' position to promote knowledge sharing, to enable the expression of synergies between local and scientific knowledge. We need to increase support to communities to develop their knowledge and institutions, and to explore complementarities between scientific and indigenous knowledge. We must promote technology transfer with respect for others' identities. We have to facilitate the control and management of lands and resources under customary arrangements, which many see as fundamental to poverty eradication, and which will require the resolution of contentious sovereignty conflicts. We must insist on peoples' informed consent to any developments in their regions, which implies greater corporate accountability. And they should maintain control of traditional knowledge and receive a fair share of any benefits accruing from its use, which implies equal status between partners.

Theory and ideology in the coming chapters

In Chapter 2 Michael Fischer reflects upon the relationship between explanation and practice, applied scientific and cultural knowledge, to argue that applications of scientific knowledge are not the same as science itself, but that these applications undergo a process that has properties not unlike those of indigenous knowledge. He asserts that this process results in knowledge that is not just about the system represented, but which is necessary for the system to operate in a contingent world; what he calls deontic or enabling knowledge. Fischer argues that describing or formalizing this enabling knowledge allows us to describe more fully experiential, informal, uncodified knowledge, the better to identify it and thereby understand how indigenous knowledge works and thus how it might be modified in a new context. His ethnographic example of a seed potato project in the Swat Valley of northern Pakistan is a salutary lesson in how the confusion of this enabling knowledge with good practice can lead to project failure.

In a similar vein, using a development example from Guatemala, Hans Siebers (Chapter 3) argues that the perceived opposition of indigenous knowledge versus modern technology leads us into serious conceptual problems. For him there are no fundamental ontological differences between the various bodies of knowledge these categories refer to. Following Hannerz, to explore this further Siebers considers how the Q'eqchi'es of Guatemala manage the flows of their own knowledge and the ways in which scientifically elaborated knowledge is transmitted to them. He describes the local agricultural knowledge specific to the Q'eqchi'es called *na'leb'* and how they deal with the flows of this and scientific knowledge. Siebers conceptualizes Q'eqchi'es management of both local and scientific knowledge flows in terms of creolization. By this he means the selective adoption and adaptation of their own knowledge flows and those that reach them through development agencies depending on whether they fit or satisfy their world views, perceptions of identity, and power relations that determine these knowledge flows. Building on this analysis Siebers draws parallels between Q'eqchi'es knowledge management and contemporary (Western?) approaches that emphasize reflexivity and experience rather than learning, and concludes that as a result intervention policies should facilitate the opening up of local knowledge and not merely its replacement.

Based on their fieldwork among the Duna people of the Lake Kopiago district in the Southern Highlands province of Papua New Guinea, Pamela Stewart and Andrew Strathern (Chapter 4) offer an ethnographic assessment of the ways in which the Duna use their indigenous knowledge of the environment and their place within their cosmological perception of the world to deal with change, particularly in the control of development projects. In so doing they conclude that if viewed historically not only are the concepts of indigenous knowledge and intellectual property rights pertinent to an understanding of contemporary development processes, but that they can be seen to provide 'popular agency' in the face of exploitation; in the case of the Duna in their negotiations with mining companies.

In Chapter 5 Colin Filer focuses on the on-going battle between the conservationist lobbies and the expansionist logging industry in Papua New Guinea. Using a dramaturgical metaphor, Filer presents the situation as public or private performances of series within the play called 'Conservation Policy', where indigenous knowledge is a sub-plot. Based on his own long-term participation in the conservation policy process in PNG, Filer tries to identify the means by which indigenous knowledge is invoked, how its role is represented by the various factions, and what impact these representations have on any resultant agreements and conservation policy. In so doing he seeks to offer guidance to, as he says, 'the debate between anthropologists who have an interest in conservation and conservationists who have an interest in anthropology'.

Drawing on ethnographic data from an Aboriginal community in North Queensland and Euro-Australian pastoralists on the surrounding cattle stations, Veronica Strang's contribution (Chapter 6) considers the relationship between systems of knowledge and attachment to land. She argues that the use of land as the primary medium for the location of cultural knowledge engenders place-based identity and affective environmental relations that are not experienced to the same degree by more transient cultural groups. Implicit in this argument is an assumption that indigenous knowledges and identity have specific characteristics and are located in place in ways that are meaningfully different to the more fluid knowledge and identity constructions of other societies.

Terence Hay-Edie's Chapter (7) describes his fieldwork in UNESCO's division of ecological sciences in Paris between 1995 and 1996. In his own words his work concerned 'the conceptualization of an inter-sectorial programme designed to revalorize forms of vernacular conservation of biodiversity based on indigenous knowledge'. This initiative provided Hay-Edie with the opportunity to explore relations beyond the UNESCO-HQ secretariat, including ethnographic research at the annual Working Group on Indigenous Populations in Geneva and a regional seminar held in India. By tracking a range of different UNESCO activities, he found numerous others like himself (but not anthropologists) as he says: 'copying and adjusting themselves to fit the broad institutional discourse of the international organisation'. His conclusion is that what come out of UNESCO-HQ are merely templates of 'predetermined global categories', but which, as his discussion of a seminar on sacred sites and biodiversity makes clear, require local input if they are to be viable.

In Chapter 8 Greg Cameron focuses on one particular Tanzanian pastoralist network (Pastoralist Indigenous Non-Governmental Organizations or PINGOs) to show how the original ideology of this network was systematically subverted by the top-down imposition of indigenous human rights from other international contexts by Western donors, and how this situation was appropriated by network leaders for their own political use – to the inevitable detriment of ordinary network members.

In Chapter 9 Ilse Köhler-Rollefson and Constance McCorkle raise the spectre that is domestic animal diversity. They take as their canvas the entire South and note that much of today's remaining diversity in domestic animal breeds survives only in traditional farming and herding communities and then only because of local indigenous knowledge and social organization. They lay the blame for this fairly and squarely on FAO and other international organizations which, they argue, have made little effort to integrate such knowledge and practice into their global strategies for understanding and maintaining domestic animal diversity. Indeed, the focus of their critique is the perceived rationale that the salvation of local/indigenous breeds lies in the gene pool these animals present that might be of potential benefit to the North or to humanity at large. The paradox and irony of the situation is readily apparent: paradoxical because, as Köhler-Rollefson and McCorkle point out, the FAO and others still do not do enough to salve the situation; and ironic because many indigenous breeds are at risk precisely because of cross-breeding policies previously promoted by these same organizations. Moreover, as the authors rightly observe, there is little or no appreciation that not only are endangered breeds usually owned by marginalized social groups – whose very survival depends directly upon these animals – but that such groups have a vital interest in their conservation.

In Chapter 10 Paul Sillitoe, Julian Barr and Mahub Alam argue from the outset that indigenous knowledge research should seek out a compatibility between local and scientific ideas to 'facilitate the targeting of development resources more effectively on the poor'. To achieve this they compare local Bengali farmers' soil classification with that of soil scientists to see where there are similarities and differences. It takes as its premises that (i) farmers' knowledge of their soils is the most relevant, and that (ii) the utilization of local soil knowledge offers potential gains over expensive land and soil surveys by placing greater emphasis on local people informing scientists rather than vice versa. Their use of GIS as a domain for integrating scientific and indigenous soil knowledge exposes the frailties of current indigenous knowledge methodology (subject of a further ASA 2000 volume, *Investigating Local Knowledge: New directions, new approaches*) and proposes indigenous knowledge-aware anthropologists as facilitators between local farmers and soil scientists in the production of 'joint maps' – maps that, they argue, would at least be more relevant to local people caught up in natural resource interventions, and thereby facilitate their participation.

Paul Spencer (Chapter 11) concludes this volume with a timely reminder of the consequences of what he so appropriately terms the tragedy of globalization. He presents a comparison of the role and import of indigenous knowledge among East African pastoralists in their pre-monetary setting with that of today. In the past, indigenous knowledge was adaptive and the means by which communities corporately sustained themselves. Subsequent changes, especially those resulting from increasing territorial confinement, have spawned variants of indigenous knowledge that favour the individual rather than the community, that are not corporately regenerated. Spencer argues that these have led to increasing inequality and patronage among pastoralist peoples and the undermining of their traditional autonomy. But, as with earlier pastoralists caught up in the spread of Islam, Spencer points out that 'this creeping process of civilization has not resolved the ecological dilemma anywhere [because] it distances those who have ultimate control over resources from the problems of sustainability'.

Note

1 We acknowledge with gratitude the advice of Paul Oldham in respect of the following discussion of the efforts of indigenous peoples' organizations to gain recognition for the rights of those they represent.

References

Agrawal, A. 1995. Dismantling the divide between indigenous and scientific knowledge. *Development and Change* 26: 413–439.

Antweiler, C. 1998. Local knowledge and local knowing: An anthropological analysis of contested 'cultural products' in the context of development. *Anthropos* 93: 469–494.

Cleveland, D.A. and D. Soleri (eds). 2002. *Farmers, Scientists and Plant Breeding: Integrating knowledge and practice.* Wallingford: CABI Publishing.

Colchester, M. 2002. Indigenous rights and the collective conscious. *Anthropology Today* 18 (1): 1–3.

Ellen, R. and H. Harris. 2000. Introduction. In *Indigenous Environmental Knowledge and Its Transformations.* (eds) R. Ellen, P. Parkes and A. Bicker. Amsterdam: Harwood Academic. 1–33.

Greenpeace International. 2002. 'Lessons of history: Stalled on the road from Rio to Johannesburg'. URL: www.greenpeace.org

Griffiths, T. 2001. 'Consolidating the Gains: Indigenous Peoples' Rights and Forest Policymaking at the UN', A Forest Peoples Programme Briefing Paper. URL: forestpeoples.gn.apc.org/UNFF/briefing (24/07/02).

Kuper, A. 1999. *Among the Anthropologists: History and context in anthropology*. London: Athlone Press.

Lâm, M.C. 2000. *At the Edge of the State: Indigenous peoples and self-determination*. Ardsley, New York: Transnational Publishers.

Malinowski, B. 1944. *A scientific theory of culture and other essays*. Chapel Hill: University of North Carolina Press.

McIntosh, I. 2002. Defining oneself, and being defined as, indigenous. *Anthropology Today* 18 (3): 23–25.

Noejovich, F. 2001. *Indigenous People in International Agreements and Organizations: A review focused on legal and institutional issues*. Social Policy Program, World Conservation Union (IUCN). URL: www.IUCN.org/themes/spg/index.html

Pottier, J., A. Bicker and P. Sillitoe (eds). 2003. *Negotiating Local Knowledge: Identity, power and situated practice in development*. London: Pluto.

Sillitoe, P., A. Bicker and J. Pottier (eds). 2002. *'Participating in Development': Approaches to indigenous knowledge*. London: Routledge (ASA Monograph No. 39).

Tebtebba Foundation. 2002. 'Indigenous Peoples Political Declaration', Prep Com IV, Indonesia, Bali 6 June 2002. URL: www.tebtebba.org (24/07/02).

World Commission on Environment and Development. 1987. *Our Common Future*. Oxford: Oxford University Press.

Chapter 2

Powerful knowledge

Applications in a cultural context

Michael Fischer

Culture and traditional knowledge are concepts developed and advanced by anthropologists over the past century or so. These have recently been appropriated and used in ways never envisioned by anthropologists, sometimes contravening the data, theory and models used by anthropologists to develop these concepts. It would be fair to say that initially most anthropologists welcomed this attention, with not a few embracing these appropriations. But there is little evidence to support many of the principal applied threads that have developed in critical politics, economic development and conservation. Anthropologists have an opportunity and an obligation to clarify and refine both concepts in the context of these (mis-)appropriations, and to clarify them for what they are: anthropological inventions used to define and enhance understanding, not to define movable property or motivate new forms of race (and racism).

In this chapter I will briefly explore some of these ideas and examples in a restricted sense intended to reflect the relationship between culture, knowledge and behaviour in a context of change, in particular with respect to the relationship between explanation and practice, the relationship between applied scientific and cultural knowledge, and application to economic development projects.

Specifically, I argue that application of scientific knowledge is not the same as science, and undergoes a process that has properties not unlike those described by Ellen and Harris (2000) for 'indigenous knowledge'. This process results in knowledge that is not just about the system represented, but is necessary for the system to operate in a contingent world even though it was not originally a subset of the knowledge being applied. This is what I call deontic knowledge, or in more familiar terms, enabling knowledge. Building on Ellen's concept of prehension (Ellen 1986, 1993), I suggest the operative principle in indigenous knowledge has similar properties. Describing or formalizing this enabling knowledge permits us to more formally describe what Ellen and Harris suggest is 'tacit, intuitive, experiential, informal, uncodified knowledge'. Using a development example from Pakistan I illustrate how confusing enabling knowledge with 'good practice' can lead to project failure. I conclude with some remarks on a 'relevant' anthropology, and the need for greater cooperation between the different strands of anthropology.

Culture

In many schools of anthropology, culture developed from a useful insight, promoted by Franz Boas and others in the early period of American anthropology, into a symbolic hydra by 1960, finally mutating into the impossible chimera we confront today.

As an anthropologist I think it unlikely that most anthropologists will abandon, or would want to abandon, the culture concept as a central component of anthropological theory, but that does not mean our concept 'culture' can remain undisturbed or that applications of the culture concept emerging within and outside of anthropology be unexamined. Re-examination has been proceeding apace for well over four decades, and although we are no closer to general acceptance of a 'core' to the culture concept, nor have compelling arguments against the culture concept emerged. In the midst of this swirl of reflection, self-reflection and contemplation, culture has actually grown more pervasive as a concept, if not a clearer one. The theme of this volume is oriented in part to how anthropology contributes to economic development. Given anthropologists' focus on the culture concept, we might expect that a part of this contribution will include use of the culture concept. But can it do the work we demand of it?

indigenous knowledge and cultural knowledge

indigenous knowledge is a term that has emerged over the last two decades to describe the knowledge of a group of people local to a given situation, sometimes used interchangeably with local knowledge (Ellen and Harris 2000: 1–2) and which I am taking to be instances of cultural knowledge. Many anthropologists have questioned the value of trying to distinguish indigenous knowledge as a special kind of knowledge (ibid: 25–26). At the core, anthropologists and practitioners have very different goals for characterizing a peoples' knowledge. Practitioners are interested in knowledge that regardless of source is enactable with respect to their own practice. Anthropologists are more interested in the knowledge itself and its interconnections with other knowledge.

Although Geertz (1966) recast and broadened Wallace's distinction as 'knowledge for' and 'knowledge of' – procedural versus declarative knowledge – there is another more primitive distinction that should be drawn: knowledge that is about the system and knowledge that is a part of the system. Knowledge about a system is knowledge, conscious or non-conscious, that addresses that system and its functioning. Knowledge that is a part of the system is bits of knowledge which must be enacted in order for the system to be, and which need not be in a form similar to Geertz's more substantive knowledge. I will call this enabling knowledge, though I have generally referred to it elsewhere as deontic, derived from deontic logic, the logic of permissions and obligations (Fischer 1992, 2002). Enabling knowledge relates to how substantive knowledge can interact or inhibit, when to shift approaches to applications, or how to proceed when information is missing. Knowledge of this variety implies an overall system that must be reproduced (in part) using this knowledge (Fischer 1994; Fischer and Finkelstein 1991). As I will amplify in the next section, it is enabling knowledge that represents many of the barriers to applications of any kind of knowledge, indigenous knowledge or not.

Ellen and Harris (2000: 4–5) present a checklist of characteristics that anthropologists and others have associated with indigenous knowledge although Ellen and Harris settle on 'traditional knowledge' as the best of an unsettling group of terms (ibid: 3). This includes attributions emphasizing the empirical, practical, applied and situated (contextual and geographical) nature of indigenous knowledge, together with aspects such as oral transmission, informality and fragmentary distribution. In their conclusion it is this latter group that forms the prototype.

> However, we believe that indigenous knowledge, in the sense of tacit, intuitive, experiential, informal, uncodified knowledge, will always be necessary and will always be generated, since, however much we come to rely on literate knowledge which has authority, has the validation of technical experts and is systematically available, there will always be an interface between this kind of expert knowledge and real world situations. It will always have to be translated and adapted to local situations and will still depend on what individuals know and reconfigure culturally, independently of formal and book knowledge.
>
> (ibid: 28)

Although I agree with the overall sense of their conclusion, their emphasis on indigenous knowledge being intuitive, informal, uncodified and oral is misplaced. These are probably accurate enough as a description of most instances of what we regard to be indigenous knowledge, but Ellen and Harris seem to take these points further, as critical to the production and use of indigenous knowledge.

They are fundamentally correct in making the point that indigenous knowledge is complex and rich in its context of application, as, I would argue, is all applied knowledge. They use Richards' (1993) account of knowledge as performance, in which Richards contests a view of knowledge as a simple list of rules and decisions. Richards notes that Hausa farmers in northern Nigeria adapt to drought by making adjustments to their cropping pattern, sowing and resowing until a secure planting is instantiated or they exhaust their resources. However these '[cropping patterns] . . . are not the outcome of a prior body of indigenous technical knowledge' (Richards 1993: 67) instead requiring interactive decision-making within a constantly changing historical context, idiosyncratic for each farmer and where that historical context constrains or directs the appropriate applications. From this Ellen and Harris suggest we should 'recognize knowledge is grounded in multiple domains, logics and epistemologies'

(2000: 18). And continuing, 'it may be far more productive to move away from the "sterile dichotomy between indigenous and western" [Agrawal (1995: 5)] which idealizes and obscures knowledge and practices, disempowering peoples and systems through artificially constrictive systems' (Ellen and Harris 2000: 18).

They continue by criticizing efforts to codify indigenous knowledge, build indigenous knowledge into policies, the politics of indigenous knowledge and the resulting diffusion of agency from these (ibid: 18–24). While this is certainly descriptive, most attempts to codify indigenous knowledge have been inadequate, attempts to employ indigenous knowledge in development projects often have mixed results, and the political structures that embed and embody knowledge mirror existing status relations. But this has no bearing on our prospects relating to more formal representations of indigenous knowledge. A better conclusion is that these states of affairs are a result of our generally poor accounts and treatment of characterizing human knowledge.

Reflecting this off their conclusions, if we are to improve our understanding of indigenous knowledge, we cannot adopt Chomsky's (1965) approach, and just look towards a 'deep structure' of indigenous knowledge, nor can we accept Richards' (1993) approach of simply accepting that each application of indigenous knowledge is an improvisational performance. We certainly should not leave the study of indigenous knowledge and cultural knowledge in general to succumb to our own 'tacit, intuitive, experiential, informal, uncodified' anthropological indigenous knowledge tradition.

Although an intuitive approach might seem attractive to some, we should not follow it for two reasons. First, given that as anthropologists we are mainly interested in knowledge that is shared to a considerable degree, indigenous knowledge is in some manner codifiable, though in a more dynamic form than we have as yet developed. It is a mistake to imagine that there is some reified version of 'the knowledge'. It is likewise a mistake to imagine that this knowledge is 'magic', only existing in the ether, or that we cannot make better efforts to avoid some of the issues that emerge from current interpretations of indigenous knowledge. If we can dynamically codify instances of such knowledge that produce results similar to indigenous agents in similar contexts, we establish that this knowledge is codifiable. It was demonstrated more than 25 years ago (Shortliffe 1976) that restricted domains of knowledge could be encoded in an expert system and enacted interactively in new contexts, in Shortliffe's case diagnosis of diseases of the blood. This work has been expanded and refined (including a number of projects by anthropologists including Benefer (1989), Furbee (1989), Behrens (1989), Read (1989), Fischer (1985) and Bharwani *et al.* (2002); see also Fischer 1994b, Chapter 8 for a review and discussion); producing expert systems for very narrow domains has been an undergraduate level project in computer science, and at least one anthropology course, for over a decade. The expert system approach has a number of drawbacks (ibid), not least that it is only descriptive. However, it does demonstrate that it is possible in principle to address in part most of the observations of Ellen and Harris.

Second, the same argument related to aggregated versus individual authority has been taking place in most disciplines concerned with people over the past two decades. Substantial advances in agent-oriented representation and modelling in computer science are beginning to be applied to the social sciences to create artificial societies (or, perhaps better, model societies) in which the properties of knowledge and its distribution can be investigated (Read 2001; Lyon 2002; Fischer 2002; Bharwani *et al.* 2002). This work is relatively new, but provides a formidable method for those who are not willing to represent and analyse their data in terms of aggregates or norms (Hobart 1993: 19).

Although unfamiliar to most anthropologists, within a few years the technology necessary to work in these terms will be accessible to most anthropologists (Fischer and Read 2001) as the requirements for computational and computer-based skills decreases.

What is needed is an expansion of the conclusions of Ellen and Harris to human knowledge and its uses, not abandonment of this study to our intuition. In particular we need to re-examine the relationship between indigenous knowledge, applications of indigenous knowledge, scientific knowledge and applications based on scientific knowledge.

Scientific knowledge, applied knowledge and indigenous knowledge

Consider the relationship between scientific knowledge, technical applications of scientific knowledge and indigenous knowledge. Scientific knowledge is derived from two gross kinds of activities. The first is the conscious examination of observed physical phenomena. This itself is comprised of establishing: (i) a class of phenomena – a classification sufficiently broad such that examples of a class appear more than once, (ii) a description of the circumstances or context under which a class of phenomena can be observed and, (iii) an account of how aspects of the context interact to create or influence the phenomena.

The second activity is more or less the converse of the first, consciously creating and manipulating a context in order to precipitate an instance consistent with a phenomena class in a replicable manner. Technical applications are derived from this second activity, but they are not science. Whereas doing science requires, in principle at least, a conscious and reflexive knowledge of the relationship between the context and the phenomena class, technical applications do not. These have different goals. Scientific application is oriented towards understanding, technical application towards doing.

Penicillin of a given dosage and frequency works equally well in the same circumstance for allopathic practitioners and *unani tib* practitioners in Lahore, regardless of their basis of understanding or explanation for how it works (Lyon 1991). At the same time, scientific knowledge is important to the engineer as a legitimating device. Knowing there is a good reason for the technology to work is apparently comforting to many practitioners, and much of engineering is involved in advancing the ritual and religion of the explanatory knowledge that underlies practical knowledge (Bourdieu 1990).

This is not to suggest that producing a technical application is simple. Applications are rarely single magic bullets. Instead, applications are created using some combination of techniques that work together for a desired result. The gross combination and sequence is often known for an application type, but detailed implementation usually requires some considerable adjustment in configuring the technology to the specific conditions of the implementation, especially in the early stages of a technology. For example, in microelectronics it takes one to two decades for a new technical development to make the transition from first implementation to wide application (Fischer 1994b). Part of this delay simply reflects the development and diffusion of knowledge relating to a new technology, but perhaps more important, it is over this time that the technology itself is refined to make it more adapted to a wider range of contexts of application by practitioners who possess less and less knowledge by incorporating accumulated knowledge of these contexts of use into the technology itself. This is similar to the pattern of development of scientific innovations, where initial demonstration of an effect often appears in a very restricted and difficult to produce context, but as the context becomes better understood, so is the effect easier to demonstrate. This process in engineering is a result of gradually describing the many contingencies that make applications difficult, and adapting the technology so that the materials, tools and techniques incorporate knowledge relating to these contingencies and thus tend to work better across the contingent range.

Technology is often a blend of knowledge about how to interact with material systems, knowledge about the interaction and knowledge about what can and cannot be done in different circumstances and how to adapt to different circumstances (deontic or enabling knowledge, usually referred to as contextual knowledge, although this usage is descriptive rather than analytic). The latter variety is more often in need of revision than the former two since the kinds of circumstances that can arise change often in contrast to underlying principles. This form of knowledge is necessary to produce results from the former two, and thus must be kept dynamically in tune with contemporary circumstances. But perhaps more significantly, without incorporation of enabling knowledge, we are in fact not importing useful knowledge at all because the powerful things that the knowledge enacts in its origin context are not present.

Development project contexts are often presented as if we are exporting techniques that are based simply on true scientific knowledge. An industry has been made of pointing out that we often do not do so. Most of this discussion has related to not exporting the context within which the knowledge must be embedded to be effective, thus not actually exporting effective knowledge. More specifically, we are not exporting useful, enactable knowledge because important contextual enablements that the exported knowledge interacts with in its origin context are not present.

Some knowledge is seen as being powerful because it is true. Scientific knowledge is often used as a case in point (though science, by definition, is contingent). But, in fact, much of our knowledge is powerful because it provides access to powerful processes and structures, not because it is in fact true. The confusion with truth comes from associating too closely philosophical truth with knowledge. For example, a knowledge of spirits cannot be shown to be true based on most empirical knowledge of the world. But a knowledge of spirits can be operative and powerful if it provides access to powerful things, powerful people or powerful social institutions.

Much knowledge that we value is thus not either Geertz/Wallace's 'for' or 'of', but is valued because it is enabling. Knowledge of this sort can include knowledge that others hold and relate to how this exterior knowledge can be enacted or how we can avoid its consequences, how to get knowledge that is suitable for a situation, and even how to simply survive until other knowledge that falls more within our conventional categories can be enacted. Much of what we deal with in the world is contingent, either because it is truly contingent, or simply because it is beyond our power to know and thus we must guess. Enabling contextual knowledge can have many parts that unfold in layers.

I characterize indigenous knowledge (in the universal sense of Ellen and Harris, not just that of indigenous peoples) as incomplete knowledge. Much of indigenous knowledge relates to accessing powerful processes (both natural and human influenced), structures and people, including the exploitation of environmental resources. Part of this access is due to conventional views of knowledge facts, classificatory systems, relationships and knowledge of processes and contexts. Another part is related to what Ellen (1993: 229–234) refers to as prehension, 'those processes which . . . give rise to particular classifications, designations and representations' (ibid: 229). In other words, those processes that Richards (1993) concludes are situational performance or improvisation. The serious study of this aspect of indigenous knowledge is required to understand how to enable a given body of substantive knowledge for applications. The value of substantive indigenous knowledge should not be underestimated as an export in its own right, but neither should it be confused with enabling knowledge, nor should we be surprised if we ignore this component and face difficulties in application. We can face problems when enabling knowledge is exported if it is inappropriate to the new contingencies within which application is desired, as Dove (2000) suggests. In either case, it is important to be able to identify enabling knowledge, both to understand how indigenous knowledge works in its original context, and how it might be modified in its new context.

An example from Pakistan

In 1981 the Kalam Integrated Development Project (KDIP) was initiated by the Swiss government in cooperation with the Pakistan Agricultural Research Council. This was intended to deal with a unified approach to forestry economy and agriculture in the upper Swat Valley, Pakistan. The programme had a very bad start (KDIP nd) and was forced to close within the first year.

One of the initial projects they had attempted was the introduction of seed potatoes to Kalam in the mountainous upper Swat Valley in Pakistan for export to the Punjab. The new crop was not well received, in part because it was a new crop and farmers did not have an immediate use for it, did not know about it, resented the insulated intrusion of the project and because the model plots were not consistent in their yield. The geographical focus of the programme, the area around and beyond Kalam, has very little level land, and most crops are cultivated on small plots terraced from the mountain sides 2,500–2,900 metres in elevation, with varying ground cover, increasingly eroded due to high levels of over-depletion of tree cover, and with a very short growing season. Significant areas of mountain arable land receive less than normal levels of rainfall, requiring irrigation from the melt from glaciers.

A local historian and religious scholar, Abdul Haq, became interested in the new plant, and set up a number of plots, taking what he took to be the best results, and using these for seed. After three years he had produced several different varieties of potato to grow in a number of different situations that are common in the region (on a level plot, a slanted plot, near trees, mixed with grass). When KDIP reopened in 1983, after consultation with the residents of the area, these were taken up by the development project, and using the descendents of these potatoes eventually went on to wide introduction of successful potato cultivation for local use, as well as plots for seed potato for export. Potatoes are now a staple crop (together with maize – an interesting story for another day) in the area.

Haq did nothing that the development project could not have done, but whereas they had focused on getting people to prepare particular kinds of plot suitable for a particular variety of potatoes, Haq had focused on adapting the potato for the land available. I talked to him about this in 1992 in the village of Bhuu: he had been inspired by traditional practices associated with sheep and goats, whereby smallholders would seek to breed their animals with animals of another smallholder with a similar type of land access. This was said to result in animals more suited for the land available. Note that this is a broad analogy on his part, since the procedures he actually used were not those of traditional breeding practice.

The other difference was in his criteria for a successful crop. KDIP had focused on how many potatoes were produced for export. Haq was interested in how reliable local crops were, and how quickly they grew, and was rather unconcerned with how big they were or gross numbers. In this process he was not using a scientific approach, though it was empirical, because he was not concerned with understanding the development of the potato or why different contexts altered the potato, but simply towards achieving particular goals. He applied knowledge in a systematic manner, but he was not, at that time at least, interested in the intrinsic validity of that knowledge, but in the application of that knowledge in a way that worked towards his goals. Because his goals were derived from his local, culture-based knowledge, he produced a result that was acceptable and accepted by others in his society as useful (after a time). It also met their criteria, and he had provided an example that made sense to them.

In 1992 KDIP was still mainly focused on production of seed potatoes, and the overall programme was identified as a success when it was wound up in 1998. They saw the Haq potatoes as a means towards an end of getting people to accept the cultivation of potatoes, potentially one of many. In 1992 they were still oriented to bigger and more potatoes for local use, and had a number of programmes to accomplish this goal, rather than examining the prospects for improving the Haq potato. In this they were not being scientific, nor were they very successful from a technical point of view with respect to local staple crops. They were, therefore, employing powerful knowledge, support for local small potatoes, that enables their main goal, the production of seed potatoes. But they persisted in attempting to change the variety of potato grown locally. They did not see the Haq potato or equivalent as enabling knowledge, but simply as one more part of the development scheme.

I was in Kalam for two weeks before meeting Haq, two more weeks before hearing his story of the potato, and two further weeks following this up with representatives of the development project, who at that time were not even aware of the origins of the current local potato crop (though Haq is acknowledged in older promotional literature for his accomplishment). They were certainly not aware of local criteria for a successful crop, because it had never occurred to them that it could be any different from their own, bigger and more. They had simply accepted the knowledge transmitted to them by their former colleagues as pro forma. It had never occurred to Haq to tell them either. He had not set out to consciously produce potatoes with particular properties, he had set out to produce potatoes that worked, and it was self evident to him when that had come about. More important, the successful cultivation of local potatoes was critical to making potatoes a cash crop for export. Discussing this issue with KIDP staff I found on the one hand they were happy to show sensitivity towards local views (doubtless necessary powerful knowledge these days), while still persisting with their views of what constituted an improvement. But a part of this failure on my part was because I was trying to get them to do too much too soon. This example is not uncharacteristic for its time. I have seen the residue of other projects in other parts of the world, such as the failed Scanwater project in Cameroon which included components unmaintainable in the local environment, and unmaintained were certain to fail (and they have). Not all organizations take this point of view: the DTZ (The German

Society for Technical Cooperation) in particular has a good record of building solutions that suit the local context and employing serious mechanisms that require that local people select and guide the projects undertaken. And, indeed, these days it is essential to use consultative processes. But effective use of these is often quite another matter because basic conceptions of improvement vary.

Conclusion

For anthropologists to make any impact on these issues, they must be able to express matters of local knowledge and culture in clearer terms, terms that work in the target culture, and for the development experts. All too often we offer them bigger and more (and more complicated) potatoes, when what they want are small, low-yielding consistent potatoes. These are technical people. They do not want to understand about culture, they want to understand how specific knowledge about local culture can help their work, and exactly how they can acquire the knowledge. They do not want to be told what we want them to know, they will not respond to critiques. They need to be told in terms they already value. Of course we should try to educate, but if this is all we do, will we fare any better than they?

Successful applied anthropologists already have made considerable advances in this direction. For their troubles, academic anthropologists often treat their work as at best substandard, more often with contempt. But applied anthropology is not only the proof of the pudding; it is the major interface between anthropology and the explicit handling of different knowledge systems. But because many applied anthropologists do not have the resources to analyse their results in a larger context, much of this work goes unnoticed by academic anthropologists. An anthropology that can do no work may be a comfort to the old guard and the new guard of academic anthropology, but it is no comfort to me.

This exclusion is probably more related to the political economy of academic anthropology than an indication of the inherent value of a working anthropology. Increasingly many anthropologists are less and less concerned with detail, and more and more concerned with possible problems that might emerge if they were to undertake detailed studies. This is exemplified by many anthropologists, who self-consciously fabricate an impressionistic edifice in favour of a scholarly argument, neatly avoiding any problems that might arise from uncritical acceptance of investigative techniques and interpretation of data by eschewing both.

I think that academic anthropology can benefit greatly from work that will assist the applied anthropologist by directing more attention to working models of culture. That is, to address the means by which a concept like culture can manifest itself in living women and men in a particular locale; how knowledge is distributed through a society and how access to that knowledge is manifest; how new knowledge is created, disseminated and reproduced; how old knowledge is retooled for the contemporary setting. This is not to say that we should put aside the current work in anthropology. Rather I am suggesting that for anthropology to advance it requires more than simply re-examining the concepts of anthropology with more and more sophisticated language, and occasionally more sophisticated insights. There must come a point where these conceptions are put to the test, evaluating how theoretical knowledge might fit into existing situations, or not.

Not that these concerns that have shaken anthropology over the past two decades were unfounded or unuseful. We had to come to grips with the fact that complex situations are complex, and that a few waves of the hand this way or that was not sufficient to reduce the complexity. We had to understand that all knowledge is contingent, not just scientific knowledge. We had to address the implications of much greater interaction between groups of people in both time and space. A focus on individuated knowledge, maintenance of agency and how groups support and reproduce knowledge, provides us with an opportunity for the discipline to begin a more unified perspective that permits anthropologists working on different strands of the problem to cooperate. That is, there is the need for detailed contextualized and finely textured studies, systematic fieldwork, development of theory, applied work and the development of new qualitative and quantitative approaches to study these issues. Ultimately knowledge, its distribution and its use, is the key to understanding humans and their influence upon the world.

References

Behrens, C.A. 1989. The scientific basis for Shipibo soil classification and land use: changes in soil-plant associations with cash cropping. *American Anthropologist* 91 (1): 83–100.

Benfer, R.A. and L. Furbee. 1989. Knowledge acquisition in the Peruvian Andes: Expert systems and anthropology. *AI Expert* 4 (11): 22–30.

Bharwani, S., M. Fischer and N. Ryan. 2002. Adaptive knowledge dynamics and emergent social phenomena: Ethnographically motivated simulations of behavioural adaptation in agroclimatic systems. *Cybernetics and Systems Research* 16. Vienna: EMCSR.

Bourdieu, P. 1990. *The Logic of Practice*. London: Polity Press.

Chomsky, N. 1965. *Aspects of the Theory of Syntax*. Cambridge: M.I.T. Press.

Dove, M.R. 2000. The lifecycle of indigenous knowledge, and the case of natural rubber production. In *Indigenous Environmental Knowledge and Its Transformations*, (eds) R.F. Ellen, P. Parkes and A. Bicker. Harwood: Amsterdam.

Ellen, R.F. 1986. Ethnobiology, cognition and the structure of prehension: some general theoretical notes. *Journal of Ethnobiology* 6 (1): 83–98.

—— 1993. *The Cultural Relations of Classification*. Cambridge, CUP.

Ellen, R.F. and H. Harris. 2000. Introduction. In *Indigenous Environmental Knowledge and Its Transformations* (eds) R.F. Ellen, P. Parkes and A. Bicker. Harwood: Amsterdam.

Fischer, M.D. 1985. Expert systems and anthropological analysis. *BICA* 4: 6–14.

—— 1994a. Modelling complexity and change: Social knowledge and social process. In *When History Accelerates: Essays on the study of rapid social change*. C. Hann. London: Athlone.

Fischer, M.D. 1994b. *Applications in Computing for Social Anthropologists.* London: Routledge.

—— 2002. Integrating anthropological approaches to the study of culture: The 'Hard' and the 'Soft'. *Cybernetics and Systems Research*, vol. 16. Vienna: EMCSR.

Fischer, M.D. and A. Finkelstein. 1991. Social knowledge representation: A case study. In *Using Computers in Qualitative Research.* N.G. Fielding and R.M. Lee. London, Sage. 119–135.

Fischer, M.D. and D. Read. 2001. *Final report to the ESRC on Ideational and Material Models.* Canterbury: CSAC.

Furbee, L. 1989. A Folk Expert System: Soils classification in the Colca Valley, Peru. *Anthropological Quarterly* 62 (2): 83–102.

Geertz, C. 1966. 'Religion as a Cultural System' in M. Banton (ed) *Anthropological Approaches to the Study of Religion.* London: Tavistock.

Hobart, M. 1993. Introduction. In *An Anthropological Critique of Development.* (ed) M. Hobart. London: Routledge.

KIDP. n.d. Kalam Integrated Development Project. http://www.infoagrar.ch/Informationcenter/mediadir.nsf/378e730c39f9de8ac12569e000393011/4b5b4e4cdc6324e1c1256517006204d4 [imprint 23 October 2001], Info-Agrar.

Lyon, S. 2002. Modelling competing contextual rules: Conflict resolution in Punjab, Pakistan. *Cybernetics and Systems Research*, vol. 16. Vienna: EMCSR.

Lyon, W. 1991. Doctors and patients: Medical practice in Lahore, Pakistan. In *Economy and Culture in Pakistan: Migrants and cities in a Muslim society.* (eds) H. Donnan and P. Werbner. London and New York: Macmillan.

McCarthy, J. 1984. Some expert systems need common sense. http://www formal.stanford.edu/jmc/someneed/someneed.html [imprint 23 October, 2001]. University of Stanford

Read, D.W. 2001. What is kinship? In *The Cultural Analysis of Kinship: The legacy of David M. Schneider.* (eds) R. Feinberg and M. Ottenheimer. Urbana and Chicago. University of Illinois Press, 78–117.

Read, D.W. and C. Behrens. 1989. Modeling folk knowledge as expert systems. *Anthropological Quarterly* 62 (3): 107.

Richards, P. 1993. Cultivation: knowledge or performance. In *An Anthropological Critique of Development.* (ed.) M. Hobart. London: Routledge.

Shortliffe, E.H. 1976. *MYCIN: Computer-Based Medical Consultations.* New York: American Elsevier.

Chapter 3

Management of knowledge and social transformation

A case study from Guatemala

Hans Siebers

The concept of indigenous knowledge has come to play a prominent role in contemporary debates on development. This coming to the fore reflects the fact that processes of social transformation and of formulating policy objectives of social intervention are increasingly understood in pluralist terms, i.e. as multiple trajectories (Helmsing 2000) or as multiple modernities (Arce and Long 2000). The emphasis on plurality indicates that in our understanding of these processes we have made a decisive step away from the conventional developmentalist ways of thinking about social transformation and intervention framed in evolutionist, teleological, ethnocentric or naive optimistic expectations. It also points to the need for highlighting the aspirations and interests of the people involved in these processes, demonstrating the value of their own resources embedded in their life-world. Social transformation can no longer be equated – if this was ever possible at all – with the adoption of modern technology, the assumed opposite of indigenous knowledge.

However, this opposition framed as indigenous knowledge versus modern technology leads us into serious conceptual problems. I would claim that there are no fundamental ontological differences between the various bodies of knowledge these categories refer to. From the instrumental point of view, the rationality of modern technology compared to indigenous or local knowledge cannot *a priori* be assumed. Modern technology is always applied in specific social conditions which determine its outcome. Consequently, the efficacy and efficiency of modern technology can only be evaluated *a posteriori*, after its application in specific conditions. As we know very well from e.g. the green revolution, the application of modern technology may easily lead to shifts in power relations neutralizing the possible positive effects of increasing rewards. Problems of instrumental rationality (efficiency and efficacy) must always be understood in relation to the specific groups of people involved the moment it becomes socially relevant. In social terms the application of a specific body of knowledge may be rational for some while being irrational for others. The efficacy of both indigenous knowledge and modern technology would adopt a fetishized character – expressed in meaningless quantitative terms – when delinked from the various groups of people involved and isolated from the cultural and power contexts in which they are always embedded.

From the point of view of power relations, however, every intention to portray indigenous knowledge as more democratic, participatory or egalitarian reflects an almost incurable naive and romantic view on indigenous people. Only in tourist guides may one expect these kinds of arguments. Also from a meaning-making or cultural perspective the terms 'indigenous' and 'exogenous', or 'traditional' and 'modern' are highly questionable when applied to specific bodies of knowledge. They may easily lead to erroneous historicist thinking. What now may seem to be very characteristic to a specific group or region may in fact stem from other parts of the world in earlier times. A nice example is presented by the raising of chickens, so very typical of local peasants and farmers in Guatemala, when in fact chicken-raising had been introduced by the Spanish colonizers. Pleas by so-called Mayan intellectuals to turn back time and eradicate all colonial influences, such as chicken-raising, do not sound very convincing for the very same reason.

We are reminded here of the fact that terms like 'indigenous', 'endogenous', 'autochthonous', and 'modern' are basically cultural constructs in need of continuous reinvention, but which have very little to say about the ontological status of the bodies of knowledge concerned. Such qualifications do not promote understanding of what various bodies of knowledge mean in everyday practices of everyday practitioners, and the processes of social transformation they are involved in. In order to understand the meaning of various bodies of knowledge I would call attention to two crucial aspects. First, the cultural context and power relations in which flows of knowledge are embedded need to be addressed. The conception of flows of knowledge (see Hannerz 1992) serves our purpose in distinguishing a plurality of knowledge influences based on classifications and distinctions the people themselves construct (emic point of view). It allows us further to demonstrate the power and culture connotations of these knowledge flows. Second, the ways in which specific groups of people deal with these flows, how they manage the various bodies of knowledge involved in their daily organizational practices (see Clegg 1994; Nuijten 1998) are crucial here.

In this chapter I want to exemplify this approach by focusing on a specific ethnic group in Guatemala, the Q'eqchi'es, and their management of various flows of knowledge. This discussion is based on my fieldwork among them in the 1990s and on my PhD thesis (Siebers 1996 and 1999). First, I will briefly introduce the Q'eqchi'es and the ways in which scientifically elaborated knowledge is transmitted to them. Second, the context in which another important source of knowledge, called *na'leb'*, is reproduced by local Q'eqchi' communities will be outlined. Third, I will present some examples of how the Q'eqchi'es deal with specific elements of both sources or flows of knowledge, referring to the use of fertilizers, maize seeds, and planting methods. Fourth, their management of various flows and bodies of knowledge will be conceptualized in terms of creolization influenced by a specific context of world views, identity constructions and power relations of the stakeholders of these knowledge flows. Fifth, based on this analysis of the Q'eqchi'es manner of dealing with these flows of knowledge, I will make some contributions to our comprehension of processes of social transformation and intervention policy drawing on the current literature on knowledge management, learning processes and policy development.

The Q'eqchi'es and scientifically elaborated knowledge

There are approximately 600,000 Q'eqchi'es living in northern Guatemala, scattered across an area of about 20,000 square kilometres. Part of this area, the heartland, consists of a mountain range whose highest peak reaches 3,000 metres above sea level. The other part is made up of lowlands that until a few decades ago were covered by tropical rain forest. The Q'eqchi'es are the primary inhabitants of this area, but a small minority of Spanish speaking Ladinos hold the local and regional power positions (e.g. they are the landlords, merchants, government employees, army officers, NGO officials, development workers, etc.). The Q'eqchi'es live in some 1,600 rural communities, including coffee and cardamom plantations called *fincas*, independent villages and cooperatives. Approximately 100 of these rural communities were destroyed by the army in the early 1980s. Some of the inhabitants were killed immediately, others were captured by the army or fled into either Mexico or the nearby mountains. Towns in the area are limited in size and number. The largest one, Cobán, has about 30,000 inhabitants, and some Q'eqchi'es have migrated to the capital or other major cities. The Q'eqchi'es have their own language, religious traditions and typical dress. In this chapter I will focus on Q'eqchi' peasants living in independent villages and on the flows of agricultural knowledge they are dealing with.

These communities are located in a rural, but not isolated area. They are increasingly involved in processes of globalization, which I would like to conceptualize in terms of global flows of people, capital, information, goods, meanings, images and technology (see Appadurai 1996; Hannerz 1992, 1996). For example scientifically elaborated knowledge on agriculture, stemming from scientific centres in mainly the US and Europe, is flowing to the Q'eqchi'es mediated by donor agencies and local development agencies, which try to convince them to adopt this knowledge.

As a consequence, the ways in which the Q'eqchi'es are confronted with such scientifically elaborated knowledge presuppose a clear Taylorist separation between those who primarily produce this knowledge, i.e. scientists in Western research centres, on the one hand and those who are supposed to consume it, in this case the Q'eqchi' small- and medium-size farmers, on the other. At the moment it is produced this knowledge is framed in a rational way. It is marked by rational claims encompassing an integrated number of unequivocal and unambiguous concepts linked by unilinear and causal relations and pretending to have a decontextualized validity, i.e. a validity not influenced by local cultural or social circumstances. Basically, its technical rationality follows the ensuing formula: 'If you have soil and climate conditions "A" and apply inputs and working methods "B", then you will have results "C".'

The most important local development agencies which offer scientifically elaborated knowledge to the Q'eqchi'es are government institutions like ICTA, DIGESA and DIGESEPE, various agencies linked to the Catholic church, and a few NGO's. In this chapter I will mainly pay attention to the transmission of agricultural knowledge by government institutions. The Instituto de Ciencia y Tecnología Agrícolas (ICTA), the Dirección General de Servicios Agrícolas (DIGESA) and the Dirección General de Servicios Pecuarios (DIGESEPE) are sub-divisions of the ministry of agriculture aiming to provide technical assistance to small- and medium-size farmers. ICTA is doing research on crop cultivation and livestock raising to determine which methods and inputs are most suited for the local natural conditions. ICTA mainly provides DIGESA and DIGESEPE with applied technology to be passed on to the local communities. DIGESA focuses on agrarian production while DIGESEPE concentrates on livestock-raising. DIGESA started to work in the Q'eqchi' region in the early 1970s but for several years, its activities were interrupted by violence. It returned in 1985 and since then, its officials have been working throughout the region. DIGESEPE has a similar network of officials, albeit on a much smaller scale. DIGESA extension workers work directly in some 85 communities where they instruct farmers on agricultural techniques such as the use of hybrid seeds, chemical fertilizers and pesticides. They teach them how to improve their food crops production allowing for a surplus to be sold on the market and how to diversify their cash crop production of coffee, cardamom, fruits, pepper and tomatoes. Vegetable production in general is promoted. The extension workers train the farmers on how to maintain the quality of the soil and the wells on their land. In order to transmit this knowledge to the farmers, each extension worker works with four or five communities which he visits once a week. He (there are no female extension workers in the region) organizes a garden in each community to demonstrate the usefulness of his technology. DIGESEPE employees work along the same lines as DIGESA. They run several schemes of livestock promotion, such as the selling of chickens at a reduced price. Their main work is focused on preventing and curing animal diseases.

Next to the DIGESA and DIGESEPE officials who themselves work in the local communities, they run a programme of *representantes agrícolas* (farmers' representatives). These *representantes* are selected by the extension workers from local Q'eqchi' community members. They receive training on the topics already mentioned, which the extension workers of DIGESA and DIGESEPE are promoting. The *representante* is supposed to work in his own community and in one or two neighbouring ones to which he is expected to pass on the knowledge he received from the extension workers. He is supposed to spend half of his time on these tasks and receives a wage.

While this knowledge passes through these agencies to local Q'eqchi' communities, it looses part of its original rational framing because it cannot avoid being marked by local social and cultural conditions. It is adapted to local circumstances by ICTA, adopted and appropriated by DIGESA and DIGESEPE extension workers and translated into Q'eqchi' language and frames of reference by *representantes agrícolas.* The role played by ICTA in this respect is mainly a technical one, i.e. selecting those aspects which are relevant in local geographic conditions from a wider body of knowledge. The appropriation by the extension workers, however, means that these knowledge aspects become part of their strategies which are influenced by their world views, their identity constructions and the power resources they have access to. While talking to many of them it became clear to me that their management of knowledge is shaped, for example, by an unconditional belief in the blessings of scientifically elaborated knowledge. They have high expectations of the possibilities of changing local economic conditions by using such knowledge that they almost perceive in a sacralized way, as a *deus ex machina* solving economic problems. These expectations are linked to a high admiration of what Western civilization has brought about. In their eyes the status and power they attribute to Western societies is radiated by the scientifically elaborated knowledge they promote.

In short, they manage this technology not just in a technical, but in a specific, culturally informed way. Here we have a remarkable paradox: that the rational and universal or decontextualized claims attached to scientifically elaborated knowledge, i.e. pretending to be valid everywhere regardless of social and cultural conditions, facilitate its adoption by agricultural extensionists in the Q'eqchi' region who associate it with particular social and cultural conditions, i.e. with the power and prestige of Western societies from which this knowledge stems. This knowledge becomes embedded in a hierarchical worldview, constructed by Ladino agricultural extensionists, with Western civilization at the top.

In line with this particular cultural embedding, this knowledge serves these agricultural extensionists as a resource in their drive to climb this social hierarchy. In practical terms this means getting a position in one of the central offices in Cobán or in the capital or to get hold of a fertile piece of land in the region to start their own *finca.* The institutional practice of these agencies means that performing well as an agricultural extensionist is not necessarily a prerequisite for reaching such goals. Being loyal to one's superior, accepting clientelistic requirements or developing good relations with local Ladino landowners is more important in this respect. Consequently, their admiration for scientifically elaborated knowledge does not necessarily motivate extensionists to see it implemented by as many peasants and farmers as possible.

This management of scientifically elaborated knowledge by DIGESA and DIGESEPE employees is quite different from the ways the *representantes agrícolas* manage this knowledge. Just like these employees they receive a wage for their work, but I have met quite a number of them who are enthusiastically trying to make the best out of the knowledge they have learned from these employees. Moreover, they consider themselves to be primarily a member of a local Q'eqchi' community, so their loyalty is first and foremost with this community. In addition, they also earn part of their living as small farmers working on their own land within the community and in practice they symbolize the possibility of combining the selective introduction of specific elements from scientifically elaborated knowledge with practices and techniques stemming from their own traditions. In short, in their practice they do not support the exclusive claims the extensionists attribute to scientifically elaborated knowledge.

Next, the way in which they are instructed by the agricultural extensionists and the ways they themselves translate this technological discourse from Spanish into Q'eqchi' language and themselves address the members of the local Q'eqchi' communities result in very piecemeal presentations of elements of this scientifically elaborated knowledge. The Q'eqchi'es come to know the various elements of this technology in several meetings without the overall coherence between these elements becoming apparent to them. Instruction meetings focus on isolated advice on how to improve production. As a consequence, scientifically elaborated knowledge loses much of its rational and coherent framing within this process of transmission to the Q'eqchi' communities.

In short, as scientifically elaborated knowledge flows from Western research centres to the local Q'eqchi' communities, it becomes embedded in the various strategies of power and culture of those who pass on this knowledge and much of its rational character is lost. This loss refers to both its rational framing and structuring which took place in the Western research centres, and to the instrumental rationalist policy claims that had been attached to this knowledge by donor agencies. To a large extent these changes shape the way this technology is managed by those who present it to the Q'eqchi'es.

The Q'eqchi'es and *na'leb'* knowledge

Scientifically elaborated knowledge is not the only source of agricultural knowledge the Q'eqchi'es have access to. The Q'eqchi'es themselves differentiate between knowledge derived from agricultural extensionists and knowledge which is transmitted to them by the elderly leaders of the local Q'eqchi' communities they belong to. It makes sense to call this latter kind of knowledge *na'leb'*, their word for 'way of knowing', 'custom' and 'advice' by someone who can be trusted.

The elderly leaders are supposed to safeguard the continuity of *na'leb'* and to instruct youngsters. They are called *pasawink*,[1] elderly couples who over the years have accumulated respect and authority by way of serving the community as *chinames*. *Chinames* are elected by the community for one or two years. Only couples, i.e. men and women who have started their own household, can be *chinames* who may be assisted by *mertomes*. *Chinames* are always organized in a hierarchical way. The first couple of *chinames* (*xb'enil*) commands the second (*xkab'il*), the second commands the third (*roxil*), and so on. In general this hierarchy consists of about five couples, sometimes assisted by up to 20 *mertomes*. A couple first becomes *mertomes* and then slowly passes through the levels in the *chinames* hierarchy to eventually become *pasawink*. *Chinames* are almost always dedicated to one or more specific saints, usually the patron saints of the community. The tasks the *chinames* execute are twofold. First, they have their obligations towards the Catholic Church. They take care of the cemetery, they keep the church building clean and decorate or repair it, and they build a new one if necessary. The second task of the *chinames* is to organize customary or traditional rituals that are performed by the community as a whole, such as the feast of the patron saint and the *mayejak* rituals. The *mayejak* rituals are celebrated at the start of the maize cycle in March or April just before clearing the land when the *pasawink* in name of the community address the local mountain and some of the 13 mountains that have a predominant position in the Q'eqchi' region. At these occasions the *chinames* perform all the organizational and practical tasks such as collecting contributions from every household, gathering the animals, candles and *copal pom* (some sort of resin) which will be sacrificed, and preparing the meals. The leading and discursive roles are played by some of the *pasawink*, such as praying in name of the community, addressing the mountains and sacrificing a turkey or part of a pig, candles and *copal pom* to the mountains. This turkey or part of a pig is buried in the 'skin' of the mountain. They thank the mountains for the previous harvests, ask them for permission to cultivate the land and as such 'open the skin' of the mountain, and to cut down all that dwells on it, ask for protection against snake bites and accidents, and finally ask for a good harvest.

The basic ideas behind customary rituals such as the *mayejak* have their origin in the fact that Q'eqchi'es live on the slopes of a mountain (*Tzuul* in their language) and in a valley (*Taq'a*). They receive their food from this mountain-and-valley (*Tzuultaq'a*). Their food crops – maize, black beans, chillies, fruits, yuca, huisquil – grow on his or her skin and their animals – turkeys, ducks, chickens, pigs – feed themselves on his or her skin. It goes without saying that the Q'eqchi'es feel very much dependent on this *Tzuultaq'a*. There is an assumption that the *Tzuultaq'a* is not a thing but a 'person' – an assumption that as such cannot be controlled or verified – and then the basic ideas of their traditional religion make sense. In their view life is reproduced in a reciprocal relation within the basic entity consisting of the mountain, the local community and their natural environment. The Q'eqchi'es have to perform the prescribed customary rituals addressing the *Tzuultaq'a* and the latter provides them with life which is inherent in the things they receive from him or her, especially maize.

By way of performing these rituals in the name of the community, these leaders take care of good relations with the mountains and all those who dwell in the universe, such as the moon, the sun, the wind, the saints and the deceased. They also encourage the community members, especially the youngsters, to respect and practice the various rules and prescriptions of how to deal with each element in the natural surroundings such as the land, seeds, plants, rivers and ponds. As such these rules (*awas*) embody an important source of agricultural knowledge which the Q'eqchi'es are supposed to practice in order to maintain good relations with the mountains on which, in the end, their harvests of food crops and the growth of their animals depend.

These relations with the mountain (*Tzuultaq'a*) are not only their primary source of *na'leb'* agricultural knowledge, they also provide them with one of their main sources of identity construction. As has been outlined above, there is a vital entity consisting of the local mountain, the rivers that come down his or her slopes, the trees, plants, crops and animals which live on these slopes and the local community which also is situated on him or her.[2] They identify themselves primarily as a member of this unit and local community. Other identifications, such as the one with the Q'eqchi' ethnic group, are clearly subordinated to this primary identification. This means that a Q'eqchi' from another village is basically considered to be an outsider.

Various units of identification point to various categories of outsiders, but the Ladinos represent the most important of these categories. The Q'eqchi'es have very good historical reasons for cherishing antagonistic feelings towards the Ladinos. I may refer to the massive violence that swept part of the Q'eqchi' region at the beginning of the 1980s as a consequence of a war mainly between two Ladino factions, the army and the guerrilla movement, killing thousands of Q'eqchi'es. I may also point to the fact that there are many stories being told about Ladino merchants cheating Q'eqchi'es and making use of the weak bargaining position of individual Q'eqchi'es peasants. I may also mention the Ladino landlords exploiting the Q'eqchi' work force on their *fincas* and provoking land conflicts with independent Q'eqchi' villages, often backed up by army violence.

Na'leb' knowledge is embedded in these identity constructions and power relations, but not only between the Q'eqchi'es and the Ladinos. It is also related to internal power relations within the communities themselves. In one of the villages I studied, a household was expelled from the community by the *pasawink* after the man had used a chainsaw in the vicinity of a cave in which the *pasawink* usually worship the mountain. Making noise there was considered to be sacrilege and thus severely punished.

Natural and chemical fertilizers

To a large extent these ways of mapping their life-world, and its social and cultural characteristics, mould both the character of their *na'leb'* knowledge and the manner in which they manage this knowledge. Before discussing in more detail their ways of managing knowledge as they may draw on both scientifically elaborated knowledge and *na'leb'* knowledge, I will present some examples of topics for which alternatives from both sources are available: natural versus chemical fertilizers, *na'leb'* versus hybrid seeds, and *na'leb'* versus profane planting methods.

First let us consider the case of chemical versus natural fertilizers. The former are communicated to the Q'eqchi'es by development agencies and may be sold to them by Ladino merchants while the knowledge of how to produce the latter stems from their own traditions as promoted by the *pasawink*. The idea behind natural fertilizers is reflected in the above-mentioned practice of sacrificing to the mountain part of the harvest of plants and crops or animals the Q'eqchi'es have received from him or her. Natural fertilizers enter into the reciprocal and cyclical relations the Q'eqchi'es have with the mountain. The aim of most development agencies to encourage the Q'eqchi'es to use chemical fertilizers is derived from a macro-economic and export-oriented perspective. It is to improve the cash crop production for export which they consider to be the best way to improve macro-economic growth.

The way the Q'eqchi'es manage this matter is clear. As such they see no problem in combining natural and chemical fertilizers. They do not consider the use of the latter as an offence to the mountain. Nevertheless, they are very hesitant to adopt chemical fertilizers. Most of them reject the use of chemical fertilizers. Only those who live close to towns and thereby have access to urban markets and those who have access to only a very small piece of land use chemical fertilizers. This rejection by most of the Q'eqchi'es is not only due to the fact that chemical fertilizers are rather expensive. It has a lot to do with the ways they identify themselves and those who offer them chemical fertilizers. As I explained above, there is a clear ethnic difference between the Q'eqchi' peasants on the one hand, and Ladino merchants and government employees on the other, and most of the former have very good reasons for not trusting the latter. As a result, most Q'eqchi'es do not want to buy chemical fertilizers because in their view this would make them too dependent on people they do not trust. Those who have direct access to towns where they can buy chemical fertilizers from various merchants and thereby can avoid becoming dependent on one of them do adopt these fertilizers.

Moreover, those who have access to only a very limited amount of land use these fertilizers if this is the only way to reach one of their basic objectives of their strategy, i.e. to produce themselves a major part of the maize they need for their own consumption.[3] The Q'eqchi'es are willing and even eager to engage in market-oriented agricultural and non-agricultural activities provided that they are able to reproduce their own food crops. The Q'eqchi'es have various reasons to do so. Of course, it seems wise for them to draw on a variety of activities in order to avoid the risks that might emerge when becoming too dependent on one single economic activity. Moreover, market-oriented activities entail entering into commercial relations with Ladinos. However, there is more. The desire to produce themselves a large part of the food crops they need is related to the reciprocal relations they are engaged in with the *Tzuultaq'a* in which their life is reproduced. Dedicating all their land to cash crops and buying the maize they need for consumption would seriously disrupt this reciprocity and thereby threaten their lives. If using chemical fertilizers is the only way to meet this objective of reproducing their food crops as well as their reciprocal relations with the *Tzuultaq'a* because they have access to only a small plot of land, they do adopt such fertilizers.

Interestingly, those who adopt these fertilizers do so for different reasons from the objectives of the agencies that promote the use of such fertilizers. These agencies are mainly interested in promoting cash crop production.

Na'leb' and hybrid seeds

On the one hand, the Q'eqchi'es have an interesting variety of maize seeds – black, yellow and white – each adapted to specific soil and climate conditions, and altitudes. This variety not only enables them to put to use the best seed in specific circumstances, but also to spread the risk of a bad harvest. On the other hand, development agencies promote the adoption of hybrid seeds in order to increase maize production and thus allow the Q'eqchi'es to sell some surplus.

Apart from the technical advantages and disadvantages of both categories of maize seeds, here again, the distrust towards Ladino traders and extensionists seriously hampers the willingness of the Q'eqchi'es to trade in their own seeds for hybrid ones. The latter have the great disadvantage that the Q'eqchi'es cannot use part of the harvest of this year as seed for the next cycle. Consequently, the adoption of hybrid seeds would make them very dependent on those who sell them these seeds, i.e. people whom they hardly trust.

However, there is also another reason in terms of worldview. As has been discussed above, in the eyes of the Q'eqchi'es their life is reproduced within a reciprocal relationship between their local community and the *Tzuultaq'a*. As long as they perform the required rituals, address the *Tzuultaq'a* and sacrifice part of the crops and animals they have received from the *Tzuultaq'a*, the latter will provide them with the things they need for their survival, such as food, which are considered to be imbued with life stemming from the *Tzuultaq'a*.

Maize is the main food crop imbued with life stemming from the *Tzuultaq'a*. As a consequence, the moment of planting maize has various ritual requirements in which the idea of life inherent in the seeds is emphasized. For example, during several weeks before this planting the couple of the household abstains from sexual intercourse because the planting – which is exclusively a male practice – is considered to be similar to sexual intercourse: inserting one's seed into the ground in order to procreate and reproduce life. Substituting maize seeds they receive from the *Tzuultaq'a* for hybrid seeds – which are not imbued with life – would seriously disrupt the relationship with the *Tzuultaq'a* and thereby directly threaten the lives of the Q'eqchi'es. I have not come across a single Q'eqchi' household which uses hybrid seeds. Their worldview directly interferes with the relationship between different kinds of knowledge and inputs in production.

Na'leb' and profane planting methods

Hybrid maize seeds are not the only example of such rejection by the Q'eqchi'es; the methods they apply to plant their maize present another. Q'eqchi'es plant their maize group-wise, i.e. a group of about 15 to 20 men gathers in the morning at the land of the household of one of them. They line up and proceed in a row while each man makes holes in the ground with a stick and puts some grains into these holes. As a result the distance between two holes and consequently two plants of maize is determined by the distance between two men. In order to increase the number of plants per square metre agricultural extensionists try to convince the Q'eqchi'es to give up this group-wise planting: to plant alone and reduce the distances between the holes.

Here, again, identification and worldview interfere. To the Q'eqchi'es planting their maize is not just a technical or economic matter, it answers to ritual requirements towards the *Tzuultaq'a*. The latter wants the Q'eqchi'es to present themselves as a community towards him or her. The group planting maize represents this community towards the *Tzuultaq'a*, today on the plot of land of one of them, tomorrow on the piece of land of another community member.

The ritual character of this planting method and the idea that maize is imbued with life are underlined by the fact that the night before the planting is done on the land of a community member, this member invites friends and neighbours to join him in a vigil. He puts the seed in front of the altar in his house, slaughters a turkey, puts some of its blood on the seed 'to feed the seed' and burns candles and *copal pom* as a sacrifice to the *Tzuultaq'a*. They address the *Tzuultaq'a* asking for a good harvest, that the maize may grow well, that no accidents may occur while planting, etc. In the early morning, the man goes to his plot of land. He puts candles in a square which symbolizes the universe. He prays and plants the first seeds. Many Q'eqchi'es put a cross on the land and bury a piece of meat, turkey soup or cocoa and burn candles and *copal pom* as a sacrifice. After that, the man joins the group of men to plant his maize. After planting, dinner is served to the group by the woman of the household.

Q'eqchi'es do not consider the act of planting maize to be 'work'. Interestingly, when the same men help the same household to collect the harvest of cash crops such as coffee or cardamom, they get paid by the man of the household. In the case of planting maize they receive no payment at all. This is due to the fact that food crops are related to the *Tzuultaq'a* whereas cash crops have no specific relevance to the *Tzuultaq'a*. Likewise, the construction of a house is usually done in a group-wise manner also without payment. The trees and plants they use as building materials have grown on the mountain so they need to address the mountain from the community in order to maintain good relations with him or her. The inauguration of a new house is a special occasion for customary rituals addressing the spirit of the house and the *Tzuultaq'a*. In short, some forms of labour are commoditized whereas other forms maintain a personalized or ritual character depending on the economic or religious meanings the Q'eqchi'es attribute to the specific labour form.

Creolization and management of knowledge

The fact that the transmission of scientifically elaborated knowledge is strongly influenced by power relations between Q'eqchi'es and Ladinos determines the outcome of many of the development projects initiated by Ladino extensionists in local Q'eqchi' communities. The general attitude of the local Q'eqchi'es is to apply the knowledge offered to them by the extensionists during the scheduled project period. In the meantime they avoid any conflict or discussion. However, the moment the extensionists stop keeping an eye on the project the villagers often stop practising what they have been told.

To the Q'eqchi'es, Ladino extensionists represent power, so any direct intervention of these extensionists has to be obeyed. The moment the direct character of this intervention disappears, the Q'eqchi'es draw back on the main power resource they have at hand, which is the relative autonomy of their local communities vis-à-vis Ladino institutions. This is the moment when the Q'eqchi'es raise questions such as whether the new knowledge serves their general purposes and whether it coincides with the basic meanings they attribute to important elements of their strategy and life-world. This is also the moment when the technical rationality of the knowledge is considered: whether it works in relation to their criteria and serves their purposes. The management of both kinds of knowledge – scientifically elaborated and *na'leb'* – by the Q'eqchi'es is shaped by the fact that both are embedded in specific worldviews, identity constructions and power relations. This may, in the end, result in the Q'eqchi'es wholeheartedly adopting the new knowledge – such as has occurred in the case of the vaccination of their animals by *representantes agrícolas* – but it may also lead to outright rejection.

Of course, there is quite a lot of variation from one Q'eqchi' community or even one household to another. However, the basic ways of managing these bodies of knowledge I would like to characterize in terms of creolization,[4] i.e. as the selective adoption and adaptation of meanings and practices that stem from their own sources on the one hand and from global flows that reach them through development agencies on the other. This selectivity means adopting some elements and adapting them to their own needs and understanding while rejecting other elements depending on the way these elements become embedded in the framework of world views, identity constructions and power relations with their historical background.

As the examples discussed above have made clear, this selectivity applies to scientifically elaborated knowledge, but I would like to stress that the same holds true regarding *na'leb'* knowledge. Not without regret in their voice several *pasawink* spoke to me about examples of *na'leb'* knowledge and practices which their parents used to adhere to and perform but which younger generations currently no longer take into account. An example of this is the idea that maize should not be sold because it has been received from the *Tzuultaq'a* and because 'it will cry' the moment it is sold to someone outside of the village. However, this is not to say that *na'leb'* knowledge has either the option of being reproduced or being left behind. In some communities I have witnessed a clear revival of *na'leb'* practices and meanings being adapted to new circumstances. Both scientifically elaborated and *na'leb'* knowledge express a dynamic character and are reinvented time after time (see Hobsbawm and Ranger 1983).

This selectivity as a basic characteristic of the ways the Q'eqchi'es manage both kinds of knowledge is greatly enhanced by two factors. First, as has been outlined above, while scientifically elaborated knowledge flows from research centres to the Q'eqchi' communities through various linkages, actors and interfaces (see Long 1992; Long and Villareal 1993), it loses a large part of its rational and coherent framing. Particular elements are presented to the Q'eqchi'es in an isolated way that helps them to select some and reject other elements. To them it is not a matter of either adopting or rejecting the whole package because the coherence which interrelates the various elements of this modern technology has been lost within the process of transmission. Second, the fact that the *representantes agrícolas* themselves selectively combine elements from both sources legitimates other community members to do the same. They represent the possibility and viability of such a selective and eclectic approach.

As a consequence of the Q'eqchi'es' selectivity, DIGESA and DIGESEPE officials are not very satisfied with the overall results of their work. They are supposed to work two or three years in each community before moving on to another one. However, the extension workers stress that it takes this time just to gain some acceptance by a community. Many of them admitted that their work is only advancing very slowly. They only work with a minority of farmers within the communities they work with. In addition, most of the extension workers told me that it is very difficult to convince the communities to apply the advice they offer.

DIGESA officials attribute the low impact of their work mainly to lack of funds, which certainly makes sense. The low prices which the Q'eqchi'es receive for new products whenever they diversify their cash crop production is another problem. Nevertheless, not only external factors are to blame. DIGESA and DIGESEPE officials conceive of their work as simply transmitting their knowledge to the communities. They show very little interest in the Q'eqchi' economic strategies and *na'leb'* knowledge and in the ways the Q'eqchi'es manage both sources of knowledge. The fact that they combine an admiration for modern technology with a clear hierarchical and clientelistic way of thinking makes it understandable that they are not interested in what really matters to the Q'eqchi'es, in their eyes those who are at the bottom of the hierarchy. They are unaware of the various considerations outlined above which explain why the Q'eqchi'es adopt or reject specific elements of the knowledge they offer them. They make no effort to look for knowledge that would complement existing Q'eqchi' strategies.

DIGESA and DIGESEPE officials do not recognize their own shortcomings in this respect. Because of their lack of interest in these matters they only see that the Q'eqchi'es adopt some of their advice and reject some without being able to understand the reasons behind the Q'eqchi'es' selectivity. Consequently, they can only blame Q'eqchi' culture for the poor result of their work. These officials point to the 'idiosyncrasy' of the Q'eqchi'es, their 'lack of education', 'the way they are', their 'superstition', the fact that almost all of them speak only Q'eqchi' and their 'fearful nature' in order to explain their reluctance to 'understand' and accept the instructions offered by DIGESA and DIGESEPE. This disparaging way of talking and thinking about the Q'eqchi'es not only explains a large part of the failure of their programmes, it also shows that differences between Ladinos and Q'eqchi'es are easily interpreted by the former in racist terms and that this racism is reproduced in closed circles without cognitive dissonance being able to penetrate or interfere.

Flows of knowledge and social transformation

My analysis of flows of knowledge has demonstrated that these flows are embedded in a wider social and cultural framework that includes power relations, worldviews and identity constructions from which they cannot be separated. For example, as scientifically elaborated knowledge flows from research centres to local Q'eqchi' communities it loses a large part of the rational framing that characterized it at the construction stage in these research centres. The moment it arrives at Q'eqchi' communities it is no longer possible to separate it from the cultural and power connotations and dimensions it has adopted while flowing. Even seemingly technical questions such as whether and to what extent it works immediately have to be extended with asking for whom and who pays the price? In socially meaningful terms raising such technical questions without mentioning to whom they refer, differentiating between various stakeholders, does not make any sense. Such a way of framing questions displays a highly fetishized character. Processes of social transformation related to knowledge management and transfer cannot be understood in such fetishized ways. They include interactions between various stakeholders with differential access to power resources and also involve overall meaning-making processes regarding oneself, the 'others' with whom one interacts and the natural and general social environment.

In principle, such a rather wide and encompassing understanding of social transformation is in line with discussions about development policies in most Western societies, international organizations and donor agencies. These discussions are not limited to technical matters such as the increase of productive capacities or the growth figures of gross national product. A human development agenda is called for which focuses on the need for the expansion of the range of social, economic and political choices of groups, and of individuals, and the securing of a decent standard of living in terms of education, nutrition and health, freedom, democracy and human security as well as sustainability (Opschoor, 1999: 3). It is about issues such as good governance, sustainability, gender relations, poverty alleviation and institution building.

However, one may wonder whether we should not understand the role played by such broad conceptions of processes of social change under the banner of 'development' basically in ideological or legitimating terms. After all, the actual policies of most donor agencies in the Western world are increasingly marked by instrumental rationality that does not fit a plausible plea for such broad objectives. To return to our subject of knowledge flows: the fuel which drives this kind of knowledge to be transported from Western laboratories and research centres to the Q'eqchi' region is provided by such donor agencies' funding. New public management concepts have become very popular among many of these donor agencies stressing the need for accountability and the strict application of rational policy instruments and methods in terms of planning and evaluation. Such concepts reinforce the instrumental rationality claims of the bodies of knowledge concerned in order to create an image of value-for-money to be consumed by their principal stakeholders, i.e. politicians, government officials and the public in general in the West. After all, these stakeholders have to be convinced of the need to continue to transfer tax payers' money or donations to these donor agencies.

These considerations suggest that institutional interests and power relations may play a more important role in the application of such an instrumental rationality discourse than the plausibility of these discourses themselves. How can such a broad conception of development objectives be forced into such a straitjacket of instrumental rationality? In my view a gross overestimation of the capabilities and viability of social intervention is required to uphold the credibility of both a broad human development agenda and instrumental rationalist claims of intervention policies.

Most of us who have fieldwork experience in so-called development projects or programmes are very well aware of the fact that such instrumental rationality claims and value-for-money images cannot stand any serious scrutiny. Such claims imply that the quality of policy itself is the basic factor determining its outcome, whereas it is obvious that any process of social transformation is necessarily influenced by a variety of often unforeseeable factors and influences. These value-for-money images ignore the very existence of unintended consequences, so inherent in any social intervention. Consequently, in order to reproduce such images closed policy formulating and evaluation methods have been elaborated which rule out the very existence of unintended consequences and hide the intricacies of the factors influencing the processes concerned.

Moreover, such instrumental rationality claims require local counterparts in the countries concerned to function according to the standards of so-called good governance. The examples of the government agencies involved in the Q'eqchi' region suggest that such requirement of good governance may be far removed from daily practice. One of the main problems here is the fact that the relations between donor agencies and local development agencies are very complicated. To be sure, the latter depend on the former in financial terms, but often this financial dependency is at least partially neutralized by a very cautious attitude on the part of donor agencies to avoid reviving images and accusations of neo-colonial or imperialist behaviour. Such cautious attitudes seriously hamper the possibilities of meeting the control, accountability and value-for-money claims of the same donor agencies.

Moreover, donor organizations may choose to focus on specific social categories, such as the rural or urban poor or women, formulating inspiring policy objectives. However, in practice they are highly dependent on the information they receive on those social categories and objectives from their local counterparts, whether they are non-governmental or official, such as in the case of DIGESA and DIGESEPE. These agencies have their own interests in receiving funding and delivering accountable results for their programmes and projects, and as such they can hardly be considered as a reliable source of information about those who are supposed to benefit from development interventions but remain hidden behind those agencies. It may cause no surprise that the construction of accountability by those involved on location may have lost most of its instrumental rationalist intentions and objectives as formulated by donor agencies.

In short, the flow of scientifically elaborated knowledge is marked by the fact that in the course of such flows and transmission the rational framing of this knowledge itself is largely lost. The same holds true regarding the instrumental rationalist policy claims that accompany these flows as formulated by donor agencies. In both cases rationality has to give way to a basically cultural and political framework in which this knowledge becomes embedded.

Management and intervention policy

After having studied the flows of knowledge and the ways the stakeholders manage these various flows of knowledge the question is raised: what can we say about interventions? How can they become beneficial or sensible to the processes of social change and the interests of those who are principally at stake, such as the Q'eqchi'es in the case presented above?

To begin with, in the analysis above, the intricacies and complicated nature of processes of social transformation in which knowledge flows and management plays a crucial role, has been shown. At the same time current policy development and organizational schemes in which both donor agencies and local counterparts are involved have been discussed. It has become clear that such schemes are inadequate to reach a trustworthy and plausible analysis of this complicated nature. As a consequence, I hold that there is a need for applied extended ethnographic research focusing both on identifying which relevant matters of social transformation are at stake in a specific group and on how these matters are embedded in natural and geographic conditions, in relevant networks of interactions and power relations, in world views and identity constructions. Such an analysis may suggest which specific interventions and which specific inputs of knowledge may compliment and support the existing ways of management of social transformation of the people concerned. Such an analysis may also lay the groundwork for tailor-made interventions based on the specific results of each research project instead of preconceived ideas and objectives of development agencies, whether they are donor agencies or local counterparts. Without such research the success or failure of every intervention becomes a matter of sheer luck or coincidence.

Contemporary management literature may also be helpful for developing a tailor-made profile of intervention for social transformation. In this respect we see an interesting shift in this literature away from strict rational conceptions of management. The classical rational management conceptions were based on the possibility of knowing and changing the social environment and internal reality in organizations in a rational way. In subsequent work on strategy and management the limits of this possibility have become acknowledged. Some have tried to uphold the idea of rationality, but within narrowly circumscribed limits and put forward the term 'bounded rationality' (March and Simon 1958; Simon 1957, 1959). Others have claimed that management and intervention are basically efforts to deal with world views, identity constructions and power relations rather than the application of pre-established schemes of rational decision-making, planning and implementation (see Watson 1994; Watson and Harris 1999). In line with this literature and with the analysis of the role played by knowledge in interventions, I hold that such interventions for social transformation cannot be conceived as just a matter of instrumental rationality and accountability.

Moreover, the subject of knowledge and ways of dealing with it is important here. Contemporary concepts of management of knowledge no longer conceive of knowledge in terms of unquestionable expertise that has to be transmitted to (ignorant) pupils or laymen in order to increase their knowledge and to improve their productivity. The Taylorist distinction between production and consumption of knowledge is increasingly questioned in the literature. The shift is from instruction to construction. The contemporary emphasis is on learning processes in which co-workers are encouraged to reflect on their own experiences and to develop their own innovative capacity rather than to apply what experts have established as the only right way to work and act. The emphasis is on learning as an open-ended process of meaning-making, experiential learning, closely related to one's own experiences rather than the adoption of rationally established expertise (see Argyris and Schön

1978; Dixon 1994; Miner and Mezias 1996; Senge 1992). Such an approach to learning fits very well to my analysis of knowledge management of the Q'eqchi'es.

In addition, if processes related to learning and knowledge transfer can no longer be understood in terms of a clear-cut distinction between experts and ignoramuses such processes can no longer be understood either in terms of 'determined', but rather as 'contingent' action. Contingent means that this kind of activity cannot be planned or foreseen, but requires analysis afterwards. As a consequence, the appropriate intervention or management approach which fits this kind of activity is not a directive, but a facilitative one (see Doorewaard and De Nijs 1998).

In this case, facilitating means supporting and encouraging learning processes of those involved and reinforcing instead of replacing their specific knowledge management, learning styles and strategies. It points to the need for studying knowledge management in the local political and cultural context in which various flows of knowledge, power and policy come together; extended ethnographic case studies. Such studies may serve to identify elements of knowledge and specific instruments that the people themselves may consider to be beneficial to their learning processes, to support their own strategies and to enlarge their space within the wider power conditions. However, the analysis of the Q'eqchi'es reminds us to the fact that – even in the case of people with hardly any power resource at hand – in the end the fate of policy depends on the willingness of the people concerned to apply it and to embrace it. Facilitating means offering them these elements of knowledge and instruments, but in the end they themselves are the ones to decide. In any case, supporting creolization cannot be done in an instrumental rationalist way that continues to characterize current policies under the banner of development.

Notes

1 In a strict sense the word, *pasawink* refers only to elderly men and the word *pasaixq* is used for elderly women, but when talking about a couple or the elderly in general the Q'eqchi'es use the word *pasawink* in a more general sense.
2 A mountain can be either male or female.
3 While using the concept of strategy I do not necessarily refer to a premeditated process of explicit and rational reflection and planning oriented towards the future but rather to those lines which with hindsight become visible in one's practices and behaviour. These lines rather emerge while one is working and acting in daily practice.
4 Originally the term creolization refers to racial categories or to the mixing of two different languages from which a third language arises. This linguistic analogy has subsequently been used to characterize cultural mixing in a more general sense and the construction of new identities drawing on various cultural sources, especially in polyethnic societies (Drummond 1980). Next, the term was applied in globalization theory pointing to the articulation of aspects stemming from the 'indigenous cultures' of social actors on the one hand and elements coming to them from global flows of meanings on the other (Hannerz 1992: 261–267). In line with Hannerz it has also

been used to depict the capacity of social actors to articulate their continuous reinvention of tradition with selectively adopted and adapted external and exogenous elements, to combine the selective continuation of pre-modern elements with selectively adopted and adapted modern aspects, within an analysis of asymmetrical power relations (Siebers 1996: 275–307).

References

Appadurai, A. 1996. *Modernity at Large: Cultural dimensions of globalization*. Minneapolis and London: University of Minnesota Press.

Arce, A. and N. Long (eds). 2000. *Anthropology, Development and Modernities*. London and New York: Routledge.

Argyris, C. and D. Schön. 1978. *Organizational Learning: A theory of action perspective*. Reading, MA Addison-Wesley.

Clegg, S. 1994. Power and institutions in the theory of organizations. In *Towards a New Theory of Organizations*. (eds) J. Hassard and M. Parker. London and New York: Routledge. 24–49.

Dixon, N. 1994. *The Organizational Learning Cycle. How we can learn collectively*. London: McGraw-Hill.

Doorewaard, H. and W. de Nijs (eds). 1998. *Organisatieontwikkeling en human resource management*. Utrecht: Lemma.

Drummond, L. 1980. The cultural continuum: A theory of intersystems. In *Man. The Journal of the Anthropological Institute* 15: 352–374.

Hannerz, U. 1992. *Cultural Complexity. Studies in the Social Organization of Meaning*. New York: Columbia University Press.

—— 1996. *Transnational Connections. Culture, people, places*. London and New York: Routledge.

Helmsing, B. 2000. *Externalities, Learning and Governance. Perspectives on local economic development*. Den Haag: ISS.

Hobsbawm, E. and T. Ranger (eds). 1983. *The Invention of Tradition*. Cambridge: Cambridge University Press.

Long, N. 1992. From paradigm lost to paradigm regained? In *Battlefields of Knowledge. The interlocking of theory and practice in social research and development*. (eds) N. Long and A. Long. London and New York: Routledge. 16–43.

Long, N. and M. Villarreal. 1993. Exploring development interfaces: From the transfer of knowledge to the transformation of meaning. In *Beyond the Impasse: New directions in development theory*. (ed.) F. Schuurman. London and New Jersey: Zed Books. 140–168.

March, J.G. and H.A. Simon. 1958. *Organizations*. New York: John Wiley.

Miner, A. and S. Mezias. 1996. Ugly duckling no more: Pasts and futures of organizational learning research. *Organization Science* 7: 88–100.

Nuijten, M. 1998. *In the Name of the Land: Organization, transnationalism, and the culture of the state in a Mexican ejido*. Wageningen: Landbouwuniversiteit.

Opschoor, H. 1999. *Sustainable Human Development in the Context of Globalization*. Paper presented at EADI General Conference: Europe and the South in the 21st Century: Challenges for Renewed Cooperation, Paris, 22–25 September 1999.

Senge, P. 1992. *The Fifth Discipline: The art and practice of the learning organization*. London: Random House.

Siebers, H. 1996. Creolization and modernization at the periphery: The case of the Q'eqchi'es of Guatemala. Nijmegen. PhD Thesis.

—— 1999. *'We are Children of the Mountain'. Creolization and modernization among the Q'eqchi'es*. Amsterdam: CEDLA.

Simon, H.A. 1957. *Models of Man*. New York: John Wiley.

—— 1959. Theories in decision-making in economics and behavioral sciences. In *American Economic Review* 49: 253–283.

Watson, T. 1994. *In Search of Management. Culture, chaos and control in managerial work*. London and New York: Routledge.

Watson, T. and P. Harris. 1999. *The Emergent Manager*. London and New Delhi: Sage.

Chapter 4

Indigenous knowledge confronts development among the Duna of Papua New Guinea

Pamela J. Stewart and Andrew Strathern

This chapter will discuss the ways in which the Duna people of the Lake Kopiago district in the Southern Highlands province of Papua New Guinea are using their indigenous knowledge of the environment and their place within their cosmological framing of the world to adjust to and cope with change. Development projects are one form of change among many that the Duna have encountered over the last 60 years. The Duna had traded with neighbouring groups from different language areas in the pre-colonial past and acquired information on other places and peoples through these interactions. The first colonial explorers to enter into the Duna area were the Fox brothers in 1934 (Schieffelin and Crittenden 1991: 97–100, 268–273). The colonial government administration was established in 1960 and missionaries started entering the area soon thereafter, bringing new ideas and new pressures for the Duna to confront. The state-introduced currency, the kina, has replaced cowrie shells and (to a lesser extent) pearl shells that previously served as wealth. Nowadays, young men earn cash by occasionally working for mining companies such as the Ok Tedi or Porgera mines. Although these men travel for work elsewhere they generally return to their home area. Frequently we have heard them say that while they were away they missed their hunting expeditions, gardens, and the place itself with its abundant forest. Their comments reflect the identification of these men with their landscape, rooted in their everyday activities.

The Duna people are horticulturalists who also heavily utilize their forest in hunting and gathering fruits and some vegetables. From the forest they obtain leaves, vines, nuts (i.e. nut *pandanus*) and fruits (e.g. *Pangium edule* and fruit *pandanus*). Marsupial meat is greatly appreciated and young men enjoy the hunting of these animals as well as participating in hunting parties that go out into the grasslands close to the Strickland river in search of wild pigs. Pigs are raised for food as well as retaining their traditionally important place in transactions, being used in compensation, bridewealth and funeral payments. The Duna groups that we specifically work with live in the Aluni valley located between the Muller and Victor Emmanuel mountain ranges. Their territory stretches down to the Strickland river and comprises at least six distinct parishes (see Stewart and Strathern 2000a, 2000b for further description of the Duna people). In addition to growing sweet potato and taro they have a variety of other crops including pumpkins, yams, maize corn, vegetable greens, bananas, fruit *pandanus* and sugar cane in garden areas. The Duna have a long-standing knowledge base of the ecological potential of their area. This environmental knowledge is strongly connected to their view of themselves within their cosmological image of the world as narrated in their origin stories (*malu*).

The Duna continue to exploit their region, which is composed of rough limestone outcroppings and sinkholes, in ways that continuously make use of environmental perturbations to the best possible advantage. For example, forest fires in 1997 during an El Niño-induced drought left large areas of forest destroyed (Stewart and Strathern 1998). In 1998 the local people were burning the remaining tree trunks left in the large sinkhole areas and clearing out the stones to make large taro gardens on the hillsides that previously had been covered by forest. They were thus able to use pockets of fertile soil that otherwise would not have been easily accessible to them.

One aspect of ecological change that is frequently commented on by the Duna is the abundance of new, vigorous grasses and ferns that in some instances have out-competed the older grasses that the people remember from their childhoods. The seeds of these new plants have blown in from elsewhere in addition to being carried in on clothing and shoes by people who have travelled to places such as the Hagen area in the Western Highlands, or Pangia in the Southern Highlands, where various grasses had been introduced as feed crops for cattle. These government-introduced cattle schemes failed for a host of reasons, including the lack of considered environmental studies prior to the start of the projects (see Strathern and Stewart 2000a for a further discussion of this point). The spread of these new grasses has not gone unnoticed and in some instances has been so recent that names have not yet been given to these plants. They are often described in terms of their physical appearance relative to extant plants or plants that had existed previously. The vast knowledge of biodiversity amongst some of our Duna collaborators is astounding. We have attempted to record as many of their ecological observations as possible as well as their comments on why they believe various environmental changes have occurred. We suggest that even though our efforts may not be at the level of a trained ecologist these sorts of knowledge should be documented in view of the rapid pace of change throughout ecosystems in New Guinea (see Sillitoe 1983, 1996 on how anthropology can effectively incorporate environmental representations).

The Duna people's keen and historically attuned observations of environmental changes reflect their long-standing ability to incorporate new crops and new situations neatly into their subsistence activities. The 1997 forest fires were followed by the spontaneous establishment of a new kind of fern on the limestone outcrops of mountain ridges beside the Aluni valley. People quickly identified this fern and called it *ayu ruku*, (today's fern). They are interested in ferns in part because some ferns are used as vegetable adjuncts to the cooking of meat and tubers in earth ovens. Some hundreds of years ago perhaps, the Duna incorporated the introduced sweet potato into their crop regimen, while retaining interest in Colocasia taro. Since colonial times they have taken to cultivating maize corn, pumpkins, chocaw, Xanthosoma taro and peanuts, blending these into their gardening cycles. Their positive acceptance of these crops reflects their awareness of each crop's potential for use or exchange and of how to fit it in with their pre-existing repertoire of crops. Two points are worth making in this regard. One is that these people's interest in environmental events is linked to their cosmology, in which there is a sense of both the gradual deterioration and the cyclical renewal of the land as a whole. Crops, fruits, and new arrivals of plants are all taken as signs of the state of the Duna cosmos. The second point is that the Duna, like other Highlanders, make a distinction between multi-crop and mono-crop gardening sites. Some newly introduced crops, such as pumpkins and maize, fit easily into multi-crop garden areas, while others, such as sweet potato and peanuts, require a mono-crop environment. The distinction between the two types of garden has enabled the Duna to experiment successfully with new crops and to allot them according to their appropriate place in the agricultural regimen. They were prepared to handle coffee, introduced as a potential cash crop by the colonial administration, in the same pragmatic way; however, lack of access to transport meant the coffee beans could not be sold, and eventually the people let the trees grow wild. They remain in old garden areas as a testament both to the people's openness to change and the ineptitude of the supposed process of development planning by the colonial power at the time.

Malu knowledge

Duna cultural ideas are marked by a particularly strong emphasis on rights of precedence over specific forms of knowledge, known as *malu* as stated above. *Malu* is knowledge that itself legitimizes claims to a definite social identity through asserting a connection with an ancestral world that includes human and non-human beings. In its most straightforward form this kind of knowledge is genealogical and consists of the recitation of agnatic steps of connection with the putative founder or founders of a named local group.

The wider membership of the local group includes cognatic kin, affines and attached visitors from other named groups. These are all notionally clustered around a core of agnates who hold prior rights over the land. The Duna recognize both agnatic and cognatic descent as a basis for local group residence and participation in community activities. The agnatic descent line of members connected with the narrative of origins forms the link with the past that provides a crucial basis for agnatic precedence within the group. Woven around the genealogy there is also a narrative of the doings of the first ancestors and how they came to establish the group and thereby its claims to a particular territory, whether its members still predominantly occupy the area or not (for examples see Stewart and Strathern 2002).

Malu narratives of this kind also notably establish transitions supposedly made from primordial times, in which the ancestors were partly spirit beings and could take animal shapes, into the times of social order as they evolved prior to the colonial interventions from the 1930s onward. These narratives therefore constitute an oral history and provide a starting point for further flexible extensions and adaptations of their details to encompass recent changes, for example the arrival of Christian missions and the more recent arrival of companies prospecting for minerals, including copper, gold, and oil or gas. *Malu* as a category of knowledge therefore further comes to intersect clearly with issues of development, because whenever such company personnel move into an area, local interest is quickened and narratives of indigenous rights to the area are marshalled.

Malu stories operate both at the intellectual level as assertions of precedence over parts of the landscape and at the practical level as tools to be used in negotiating with the outside world. In a broader sense again the term *malu* refers to any kind of knowledge which for the people is authenticated through intergenerational transmission. If, for example, a version of a *malu* narrative on any subject is contradicted, the narrator is likely to say, 'This is the *malu* as I heard it from my father, perhaps this person's father gave them a different *malu*.' Different versions are thus accorded a nominally legitimate status; but the narrator tends to stick with his own version, since that is his *malu*. There is thus both a recognition of the relativity of knowledge, and an assertion of the primacy of the narrator's own version of that knowledge for his kin and place. The concept extends also to different forms of knowledge such as kinds of plant medicines or stories about parts of the landscape where particular events took place in the past. This approach to knowledge can accommodate pluralistic versions and provides over time a flexible repertoire of information without attempting to synthesize it systematically between people. While in terms of its own assertion of authenticity through ancestral transmission this form of knowledge might appear to be rigid and compartmentalized, in practice *malu* stories are at least partially open to variation and alteration over time, and since *malu* are extremely detailed and can be very lengthy, various aspects of the narrative can be truncated or emphasized depending on the particular context at hand. Thus, *malu* are an important part of the people's overall adaptation to historical changes that confront them. They act as a means whereby external changes are internalized and are made to appear as a part, even if an awkward part, of the Duna cosmos.

At the broadest level *malu* knowledge provides an explanation for the people of how things have come to be as they are. It also gives them the practical means to handle this explanation by prescribing sacrifices needed to keep the spirits of the cosmos reasonably benevolent towards humans. Most recently, as we will show, *malu* knowledge has entered into the people's reaction to drilling activities in their area and has provided them with an organizing principle for seeking compensation for these activities.

Malu knowledge being contextual, each narration, or use of an account, is particular to the occasion involved. In practice many people may know parts of a given *malu* narrative through hearsay from others or through some extended kin ties of their own, but they are always quick to declare whether they are its true owners or not; that is, whether they can use such a narrative to claim rights for themselves or not. The telling of a story, or the fact that it may be told, does not therefore in itself constitute a claim to it as one's own *malu*, nor does it infringe the rights of those who do claim it as their *malu*. On the other hand, of course, the true owners of a *malu* must have good knowledge of it in order to validate their status. Transmission of *malu* knowledge to its proper custodians is thus important and takes place over many years. It is this process which has been put at risk since the 1960s with the arrival of government officers, schools, missionaries, and migrant labour. Some men are much better at storing *malu* knowledge and narrating it than others. These men also tend to come to the fore in contemporary interactions with mining companies.

Recent changes

During the 1990s, however, and through to the present, the impingement of processes of economic development especially among the Duna themselves but also in neighbouring areas, such as those where the large Ok Tedi and Porgera gold and copper mines are located, interestingly brought with it a resurgence in the importance of *malu* knowledge and re-stimulated its transmission from the senior to younger generations. Throughout this far western region of the Papua New Guinea Highlands, mythological ideas about spirits of the ground have become entwined with perceptions of the sites where development activities, especially mining and prospecting for minerals, take place.

For the Duna and the nearby Paiela (Ipili speakers) and Huli these mythological ideas tend to be centred on the notion that a giant python guards the resources sought by miners and resents the extraction of ore from the ground (Ballard 1998; Biersack 1999).

Ungutip, Wabis and Sillitoe (1999: 63) note that this idea is found also among the Wola people of the Southern Highlands, writing that 'rumour has it that the oil and gas deposits [which are sought throughout the Province by drilling operations] are the urine and excreta of an enormous, seven-headed snake which dwells underground'. They note that this theme may reflect to some degree on adoption of Huli cosmology. The way in which such cosmological ideas are converted into a narrative of recent history is also shown in the Wola story that the oil-excreting serpent fled to Papua New Guinea following fighting between Israel and Iraq. Technological themes are further woven in through an image of the company's computer aiming explosive arrows at the serpent's seven heads. If the company misses one head the beast will rise up and make the world turn over (loc. cit.). Cosmology, world news, computer technology, and biblical ideas of the apocalypse are all linked together here. The date predicted for the snake's uprising was the year 2000. The image of the seven-headed snake appears to be derived from Revelation 13.1 in which the Beast that is the Antichrist is envisaged as rising out of the sea, having ten horns and seven heads. The indigenous appropriation of this theme suggests that the beast may be the rightful original 'owner' of mineral deposits and cannot be regarded as simply an 'evil' force.

Another idea is that the minerals are actually also the possessions of a female spirit who can appear in human form known to the Duna as the Payame Ima (Strathern and Stewart 2000b). One Duna story (apparently of recent provenance) tells how a pair of such female spirits, called Papumi and Lupumi, left the Duna area and went to 'sit down' on top of the mountain Oko Mamo, the mountain where the Ok Tedi gold mine was later established (Stewart and Strathern 1997; compare also Ungutip, Wabis and Sillitoe loc. cit. on the spirit Horwar Saliyn). The narrative establishes a link, at least of a notional kind, between the Duna and the gold mine, helping them to feel that they have a stake in it, an idea that fits with their occasional employment there. This is an example of a mode of thinking and knowledge construction that came into play very strongly during 1999 when an oil company established an exploratory drilling rig just across the Strickland River at the far western end of the Duna area, prompting the local residential groups to announce mythologically-based claims on the rig site.

These claims were all presented as integral parts of pre-existing *malu* narratives, perhaps elaborated and made more specific because of the novel presence of the oil company but not entirely manufactured through recent invention. Younger men who previously had only an imprecise knowledge of the details of these *malu* now sought to acquire a more comprehensive understanding of the narratives. Senior men were willing to impart some of the details in order to empower their group as a whole to receive anticipated payments for environmental damage and for royalties. The indigenous knowledge corpus was therefore given a boost by the new development activities. At the same time there was a potential for conflict over rival narratives between groups, although there was considerable consensus on which groups were primarily involved.

In the event, however, oil was not discovered during 1999. The local Duna people were undeterred in their view that a huge amount of oil, which they described as a 'lake', was actually there. They developed a story that explained why the drillers had not been able to find it. In the technical account of the oil company the drillers had met solid rock at a certain depth and were unable to break through it after several attempts that left them with broken drilling bits. In the people's story this event was given a political and moral aspect. What had happened, they said, was that a local boy of one of the clans who claim land-ownership rights in the area nearby had been taken underground by two spirit women (*Payame Ima*) and there he had met a huge old man in a town made entirely of paper money. This giant was the land-owning spirit (*Tindi Auwene*) and the drilling bit of the company was pointed directly at his heart. Other parts of his body already had holes in them, representing other mine sites in Papua New Guinea. He asked the boy to take a metal pipe and knock the drilling bit away or break it. The boy did so and saved the life of the *Tindi Auwene*. This male spirit then sent the boy back to his people with a number of presents, including some money to give to his mother, some cans of beer, a very large tee-shirt, and some frozen chicken meat (all introduced goods).

This story re-establishes an image of indigenous agency in the face of development. The Duna were not, however, against the development project as such. Rather, they wanted the drilling site to be resituated on their own side of the Strickland River, to which they held more secure claims than they had on the far side where the rig had been placed. They argued that the original location had been chosen wrongly. If the drilling operation had been placed on their side and oil had been extracted they thought that this would have depleted part of the *Tindi Auwene*'s substance but not killed him. This removal of part of the land-owning spirit is what would have to be compensated for by the company to the people. The story also validates in particular the claims of the boy's own clan to the area.

Complex arguments between groups often arise in contexts like this where compensation payments are in the offing, and people reorganize their indigenous knowledge in order to steer the benefits of development in their own direction, either as individuals or for their kin groups within the wider community. In this part of the Duna area many of the senior men who hold the most detailed *malu* knowledge do not speak or understand well the lingua franca Tok Pisin, still less English, and they feel at a disadvantage when faced by company personnel who speak only English or whose Tok Pisin is too fast and is inflected with a coastal accent rather than a highland one. Under these circumstances senior men depute younger men of their group with different language skills from themselves to act as spokespersons. They are reluctant, however, to hand over all their knowledge to these younger men because to do so would mean losing their own power and control over community events. They prefer a gradual transmission. The younger spokesmen in turn may claim more knowledge than they have, fraternize across clan lines, and also seek their own personal advantage. indigenous knowledge in this context becomes a contested and ambiguous area in which struggles for power take place.

In general this example shows how people's interpretative stories of company activities not only provide for them a framework of understanding that relates such activities to their own world but also, in doing so, give them a means of negotiating with the company for things that they want. They do not see drilling as a simple technical act. Rather, by investing it with their own narrative meanings they are able to deal with it socially and politically and to exert a certain amount of influence over it. We may call this the strategic use of myth. Magic and ritual may similarly be brought into play for strategic purposes. When we speak of indigenous knowledge we tend to refer to what we understand as technical knowledge only, not myth or magic. Yet indigenous technical knowledge cannot confront western technical knowledge on its own terms. By shifting to the realm of myth people seek to re-establish their own agency.

For the people at large much more is at stake than the immediate payments of compensation. They have come to expect long-term advantages, such as the building of roads, schools and clinics, to flow from mining company activities of mining. The knowledge that is deployed to lay claim to these benefits is precisely specialised *malu* indigenous knowledge. From a situation in which it was devalued in the face of Christian missionary teaching and government secular education, *malu* made a return into social life as a valuable source of power and attention. Little wonder then that it is also creatively changing and has become a focus of contested claims between groups. indigenous knowledge in this sense becomes an integral part of the development process, and a tool by means of which the Duna, like other Highlanders, bargain for compensation payments with companies and counter the companies' technological superiority and hegemony.

Discussion

In approaching the question of what indigenous knowledge is, anthropologists have moved from a straightforward taxonomic scheme as exemplified in the ethnoscience work of the 1950s and 1960s to a more nuanced view of knowledge as historically shifting and fluid, negotiated through time in interactions with others. The latter viewpoint does not seek to deny the existence or importance at a given moment in time of taxonomic classifications that people themselves may make; and indeed such classifications may persist over time also. Furthermore, not all knowledge even of the natural world is taxonomically structured, and the content and connotations of a particular cognitive category can shift. The recognition of historical change in indigenous knowledge systems is a significant prerequisite to understanding how such systems interact with the agencies involved in the work of economic development. It is evident, therefore, that ethnographic studies of indigenous knowledge cannot confine themselves to a single point in time but must themselves be conducted historically. In particular, if a development programme is in hand studies of how indigenous knowledge and ideas alter in connection with this process need to be made on an open-ended basis.

It is clear from the foregoing ethnography that *malu* is an important cognitive category for the Duna. We could also arrange *malu* taxonomically in relation to other categories such as *hapiapo*, 'stories of things that happened in the past'. Not all *hapiapo* are *malu*, because many are narratives that are not tied into a specific ancestral charter validating claims to resources. *Malu* are therefore a special case of *hapiapo*. Rights over the assertion of claims via the recitation of *malu* are restricted, whereas rights over *hapiapo* that are not *malu* are more fluid. However, the boundaries between these forms of knowledge are not rigid. In circumstances of change we find that people may appropriate parts of *hapiapo* stories, or parts of *pikono* (sung ballads on themes from *hapiapo*, see Strathern and Stewart 1997; Stewart and Strathern 2000b) into refurbished *malu* narratives. And conversely parts of old *malu* may over time become *hapiapo*. In the case of the oil-rig on the Strickland and the story that developed around it, themes were taken from the sacred *malu* of particular groups and woven into a narrative of a balladic kind which functioned at two different social levels. At one level the story validated the claims of a particular local group; while at another level this new narrative also made certain claims on behalf of the Duna groups as a whole vis-à-vis their neighbours across the Strickland and also in relation to the foreigners who were setting up the actual rig. The story was therefore a *malu* in the making, dealing with a new context of historical experience.

This example of *malu* draws attention to the flexible mythological and symbol-making capacities of indigenous knowledge. The same capacities for change show in other contexts. In practical terms the Duna, like other New Guinea Highlands peoples, have proved themselves very adept at taking up new plants and gardening practices, as we have already noted. As one example, they learned from the Huli people, who live to their southeast and who were brought under administrative rule somewhat earlier than themselves, about the planting of peanuts as a new crop. The Duna planted this crop in circular ridges with a sunken area of ground at the centre. They did this to provide adequate drainage for the plants while also keeping moisture in the ground generally. This gardening form was an inversion of the ordinary process of making mounds for planting sweet potato, with drainage spaces around each mound. Modified sweet potato mounds were made into peanut ridges. This example shows clearly how indigenous peoples such as the Duna rapidly assimilate innovations proceeding from some geographical centre of dispersal and are able to incorporate these neatly and successfully into their existing gardening regimens, without any form of external monitoring or advice. Indeed these kinds of adaptations are often among the more successful of innovations, since they proceed experimentally and voluntarily by trial and error. We may suppose that this was also the predominant pattern of agricultural change in the past. And the example underlines the point that when change is introduced under the banner of development, close attention needs to be paid by development agencies to the people's own technical knowledge, just as attention needs to be paid to the cosmological frameworks within which technical knowledge is deployed in social action.

Another example of the adaptation of new ideas from outsiders can be found in the burial practices of the Duna. The Christian missionaries and government officials told the local people that they needed to bury their dead in the ground. They said that this was more hygienic, reducing the spread of various diseases – a point which the Duna contest. Previously, the dead had been exposed on burial platforms until the desiccated remains of the dead person were ready for secondary burial within limestone caves in the high forest. Thus, burial in wooden coffins was begun but the new practice was adapted to accommodate

Duna ideas of substance transfer and cosmological placement (see Stewart and Strathern 2001). The Duna believe that the 'grease' (exuded bodily fluids) of their dead must be reabsorbed directly into the soil to ensure the fertility of the land and to sustain the continuity of the Duna people according to their cosmological ideas which embed them firmly within their specific environment. Thus, the coffins used for burial have openings in them to allow the grease of the deceased person to flow out into the soil. The coffin is also placed on a platform, within the sunken grave, architecturally reminiscent of the platforms used previously in platform burials but now put under ground. In addition to this, it is reported that the deceased's bones may be exhumed and removed to limestone caves for secondary burial after an appropriate interval of time has passed.

The example demonstrates how the Duna can take practices that are sometimes imposed upon them and ingeniously transform them so that they still correspond to their notions of proper custom. We have been told that the oil that is thought to exist in the Duna area is in part the transformed product of the 'grease' of previous generations of Duna and that if all of the oil were extracted from the ground the soil would become dry and lose its fertility (see again Ungutip, Wabis and Sillitoe 1999: 63). The Duna have said that if oil is removed from their territory they expect to receive compensation from the companies involved because of the subsequent environmental impact that would result from this depletion. This idea runs parallel to ideas about the *Tindi Auwene* given above.

In general, anthropologists can continue to play a useful role in the context of development work by stressing their strongest points: first, as we have argued here, that indigenous knowledge is both complex and fluid; and second that the success or failure of development schemes depends on a large number of intersecting factors that go beyond economic calculation and immediate environmental impact, the two main factors currently favoured in international development planning. Whatever theoretical frameworks or methodological orientations are adopted, these basic points remain incontrovertible. Economic and environmental calculations therefore need to take into account what anthropologists have to say about the 'imponderabilia' of development processes. Moreover short-term and long-term economic considerations can themselves be sharply divergent.

Many examples could be adduced to support these points. They are the stock in trade of anthropological observation, and they remain also a significant form of intellectual capital in debates about development. In Pangia, in the Southern Highland province of Papua New Guinea, as mentioned earlier, the colonial administration set up in the 1960s a number of cattle schemes intended eventually to bring in revenue for local communities. Unfortunately, the schemes were set up in too communalistic a way, resulting in ambiguities over who was responsible for the care of the cattle. In practice, people tended to claim the cattle individually, and they wanted to dispose of them individually also. This did not suit the colonial officials who were trying to keep control of the sales or slaughter of the cattle and to deposit cash in bank passbooks. Moreover, the cattle used up large areas of rough pasture and made it muddier and rougher with their hooves, thus destroying areas previously used for gardens. Further, the cattle were sometimes poisoned by eating a weed (called *wiringou* in the Pangia language) that grew in these rough areas. The schemes tended to fail over time. The local people had classified the cattle as kinds of pigs (*kai*) and had applied their own ideas about their sale or exchange accordingly. Curiously, these cattle schemes were themselves the product of a World Bank report of 1964 that recommended that cattle be introduced rather than building projects on pig herding, because pigs were too closely bound up with social and ritual life! The development planners did not anticipate that the New Guineans would assimilate cattle to their own category of pigs, thus immediately subverting the World Bank's own odd logic.

The eventual costs of restoring areas to ecological good health after some development projects have run their course are often very large. In the Pangia case the ecosystem was not, as far as we know, returned to its earlier state, although over time abandoned cattle project enclosures would see a regeneration of fallow trees and a long-term ecological recovery. In the case of the Ok Tedi gold mine in the Western province of Papua New Guinea, the long-term effects of riverine pollution from industrial tailings were finally recognized as very severe in the year 2000 and it was unclear if the company involved, or any governmental agency, could in any way meet the enormous costs of reversing or ameliorating the effects of the pollution on down-river settlements. Anthropologists and ecologists can combine in contexts like these to temper short-term economic calculations by insisting that the people's own agency and the long-term future of the environment should be taken into account.

From another viewpoint our findings here can be compared with some of the concerns that emerge from the corpus of writings on questions of indigenous knowledge and intellectual property rights. Problems with the concept of indigenous knowledge have to do with the definition of what is indigenous, what is meant by knowledge as distinct from performance, and what is to be included or excluded as a part of a body of knowledge (Antweiler 1998; Brush and Stabinsky 1996; Richards 1993; Semali and Kincheloe 1999). With regard to intellectual property rights, similar problems have been raised as to whether such a concept is itself a Western imposition on indigenous notions. In spite of such major difficulties, the concepts do prove to be applicable to our ethnographic case, but only if they are adapted to fit with the materials themselves. Most of the writings on indigenous knowledge and rights over it have to do with genres of technical knowledge and how these enter directly into development projects, usually those having to do with agricultural improvements. In this regard, a secondary purpose of our argument has been to support the viewpoint that indigenous knowledge among the Duna is generally highly empirical and adaptive. Our primary focus, however, has been on a very different sector of knowledge, centred on the *malu* narratives that validate claims to the landscape at large and come into play, in a new way, in negotiations between companies drilling for minerals and the people themselves. Our example shows that these *malu* acquire a renewed vitality and significance in this context and are directly important for arguments about compensation payments. At the same time, in a broader sense they enable the people to feel they understand, and to some extent can control, the activities of the companies. This indigenous knowledge is therefore a vehicle for popular agency in a changed contemporary context. The example therefore further demonstrates the need to take a broad definitional line in relation to knowledge, and to see that intellectual rights are bound up with deeply practical issues that link local cosmology with negotiations over monetary compensation payments. The restricted rights over *malu* stories that protected group claims before are turned into claims for payments from commercial enterprises that have entered the Duna area. In these senses the concepts of both indigenous knowledge and intellectual property rights have proved themselves to be highly relevant for understanding contemporary development processes.

References

Apffell-Marglin, F. and S.A. Marglin. 1990. *Dominating Knowledge: Development, culture and resistance.* Oxford: Oxford University Press.

Antweiler, C. 1998. Local knowledge and local knowing. An anthropological analysis of contested 'cultural products' in the context of development. *Anthropos* 93: 469–494.

Ballard, C. 1998. The sun by night: *Huli* moral topography and myths of a time of darkness. In *Fluid Ontologies.* (eds) L.R. Goldman and C. Ballard. Westport, CT: Bergin and Garvey. 67–86.

Biersack, A. 1999. The Mount Kare python and his gold: Totemism and ecology in the Papua New Guinea highlands. *American Anthropologist* 101: 68–87.

Brush, S.B. and D. Stabinsky (eds). 1996. *Valuing Local Knowledge: Indigenous people and intellectual property rights.* Washington, DC: Island Press.

Escobar, A. 1995. *Encountering Development. The making and unmaking of the third world.* Princeton: Princeton University Press.

Richards, P. 1993. Cultivation: Knowledge or performance? In *An Anthropological Critique of Development. The growth of ignorance.* (ed.) M. Hobart. London: Routledge. 60–78.

Schieffelin, E.L. and R. Crittenden (eds). 1991. *Like People You See in a Dream. First contact in six Papuan societies.* Stanford, CA: Stanford University Press.

Semali, L.M. and J.L. Kincheloe. 1999. *What is Indigenous Knowledge? Voices from the academy.* New York: Falmer Press.

Sillitoe, P. 1983. *Roots of the Earth. Crops in the highlands of Papua New Guinea.* Manchester: Manchester University Press.

—— 1996. *A Place Against Time.* Amsterdam: Harwood Academic Publications.

—— 2000. Let them eat cake. Indigenous knowledge, science and the 'poorest of the poor'. *Anthropology Today* 16 (6): 3–7.

Stewart, P.J. and A. Strathern. 1997. Sorcery and sickness: Spatial and temporal movements in Papua New Guinea and Australia. Townsville, Australia: James Cook University, *Centre for Pacific Studies Discussion Papers Series* 1: 1–27.

—— 1998. Papua New Guinea a year after the drought. Environmental health issues in

the Aluni Valley in the Duna area. *Avenir des Peuples des Forets Tropicales Newsletter* 18.
—— 2000a. Speaking for life and death: Warfare and compensation among the Duna of Papua New Guinea. National Museum of Ethnology, Osaka, Japan. *Senri Ethnological Reports* 13.
—— 2000b. Naming Places: Duna Evocations of Landscape in Papua New Guinea. *People and Culture in Oceania* 16: 87–107.
—— 2001. *Humors and substances: Ideas of the body in New Guinea.* Westport, Connecticut and London, England: Bergin and Garvey
—— 2002. *Remaking the world: Myth, mining and ritual change among the Duna of Papua New Guinea.* Washington, D.C. and London: Smithsonian Institution Press.
Strathern, A. and P.J. Stewart. 1997. Ballads and Popular Performance Art in Papua New Guinea and Scotland. *Centre for Pacific Studies Discussion Papers* 2: 1–17. Townsville, Australia: James Cook University.
—— 2000a. *Arrow Talk: Transaction, transition, and contradiction in New Guinea Highlands history.* Kent, OH and London: Kent State University Press.
—— 2000b. *The Python's Back: Pathways of comparison between Indonesia and Melanesia.* Westport, CT and London: Bergin and Garvey.
Ungutip, W. Wabis and P. Sillitoe. 1999. Some Wola thoughts on the year 2000. In *Expecting the Day of Wrath. Versions of the millennium in Papua New Guinea.* (ed.) C. Kocher Schmid. Port Moresby: The National Research Institute. 57–69.
Warren, D.M., J. Slikkerveer and D.W. Brokensha (eds). 1995. *The Cultural Dimension of Development: Indigenous knowledge systems.* London: Intermediate Technology Publications.

Chapter 5

The knowledge of indigenous desire

Disintegrating conservation and development in Papua New Guinea

Colin Filer

Talk of indigenous knowledge

How can Western conservationists talk to Melanesian landowners about 'indigenous knowledge', when Melanesian landowners do not think of themselves as 'indigenous people', and would rather talk about that Western form of knowledge that commands the gateway to 'development'? How could Jared Diamond (1997) answer Yali's question in a way that would make Yali think again about the value of the knowledge that had failed to yield the 'cargo'? Such is the question which hangs over the battle for 'biodiversity values' in the lowland rainforests of Papua New Guinea (PNG). At the level of policy, this is a battle between a group of Occidental donors, led by the World Bank, and a group of Oriental loggers, led by Rimbunan Hijau. The choices made by Papua New Guineans, both at the level of the state and at the level of the village, have become the prizes in this tug of war.

Over the past decade, the PNG government has been persuaded to impose severe restrictions on the further expansion of the logging industry, but these policy measures have not stopped the loggers from making promises of development to local landowners whose forests have not yet been logged, and they have not enhanced the government's own capacity to persuade the landowners to stop listening to such promises. So this latter task has been left to Western conservationists, whose projects are thus designed to win back the hearts and minds of landowners who tend to blame their own government for their own lack of development. The landowners count as landowners because the point at issue here is the use of customary land, on which more than 99 per cent of the country's natural forests happen to grow. Indeed, they have come to think of themselves as landowners, or sometimes as resource owners, because they have come to believe that their best chance of development is to sell the natural resources that their land contains (Filer 1997). But their determination to defend their customary title to the land itself has also grown apace, because they do not trust the government to do so, and because its loss would leave them powerless and impoverished (Ballard 1997). Their land is their last card in the gamble for modernity, and their status as landowners is critical to their self-esteem, if not to their standard of living.

The broad political setting of the conservation business in PNG can thus be represented as a dialogue between four parties, where the voice of the PNG government is aligned with the global interest of the donor community, but where landowners (and their political leaders) have the upper hand in deciding the voice to which they will listen, and where conservationists must find a path between the empty promises of both the government and the developers to get their message across to the custodians of biodiversity (Figure 5.1). In this context, it is understandable that conservationists have been attracted to the global fashion for 'integrated conservation and development projects' (ICDPs), recognizing that they may well have to make some promise of development to local communities in order to compensate for the government's inability to enforce the conservation option. On the other hand, the prevalence of customary tenure makes it difficult or impossible to establish buffer zones around protected areas, of the kind which have been advocated by proponents of ICDPs in other parts of the world. In other words, the trade-off between conservation and development cannot simply be construed as a sort of land use planning exercise, with or without local participation, but has to be adapted to the social realities of the Melanesian landscape.

In this context, we might say that local landowners or resource owners also count as indigenous people, even if they do not use a comparable phrase themselves, because such people are normally defined, at least in part, by their 'close attachment to ancestral territories and to the natural resources in these areas' (World Bank 1991). But most definitions of indigenous peoples also place some emphasis on their subordinate political status, as ethnic or tribal minorities which are distinct from the dominant society, and the concept of a 'dominant society' does not help us to understand the role of landowners, or for that matter the role of national politicians, bureaucrats and conservationists – who are also landowners – as arbiters in a tussle between Occidental donors and

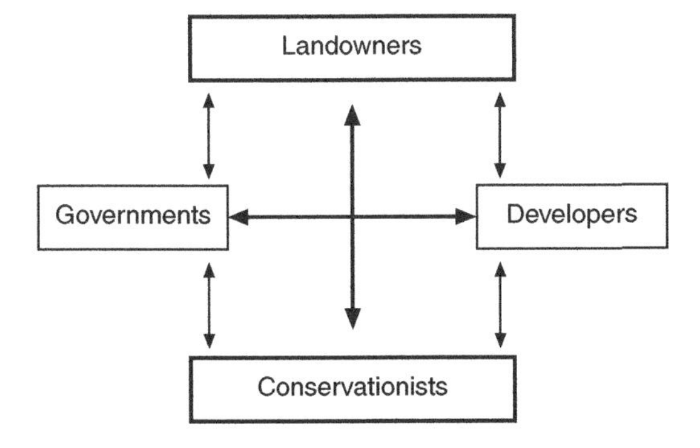

Figure 5.1 The central axis of stakeholder relationships in the PNG conservation business.

Oriental loggers. Furthermore, the idea of a 'close attachment to ancestral territories' can all too easily beg one of the main questions which I wish to pose in my account of this struggle, by leading us to assume that these indigenous people are also 'rainforest people', with a traditional interest in the conservation of natural forests or forest species.

This is where the subject of 'indigenous knowledge' enters the picture, as the topic of a dialogue within a dialogue, a type of scene or sub-plot in the public or private performance of conservation policy. It might be argued that the value and scope of indigenous knowledge has already been diminished or distorted once local actors assume the character of landowners or villagers, rather than indigenous people. But we do not have to assume that all indigenous knowledge belongs to indigenous people, that indigenous knowledge is necessarily distinct from other forms of knowledge, or that any form of knowledge is the special property of an equivalent class of people. The semantic point at issue is connected to the economics and politics of conservation in a specific regional and sectoral policy domain.

A lot of talk about the value of indigenous knowledge takes place when the pronouncements of government policy and the blandishments of logging companies are both out of earshot, on the other side of the battlefield, and members of the conservation community are divided by the roles that they play in the act of communication with local landowners. On one side are those Melanesian conservationists who are employed as the frontline foot soldiers of donor-funded conservation projects. On the other side are those Western anthropologists who may appear as listeners or participants in the frontline conversation, as passive scientists or active consultants, and sometimes even as project managers or technical advisers. And behind them are those Western conservationists who are not anthropologists, and whose ultimate aim is to address the needs of nature rather than the aspirations of its local guardians (Figure 5.2).

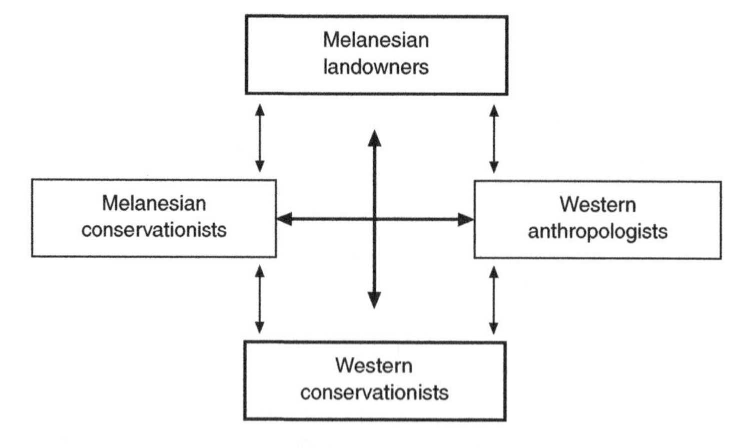

Figure 5.2 The characters engaged in talk about the value of indigenous knowledge.

How does the topic of indigenous knowledge actually enter the construction of this play within a play, and how does its appearance affect the construction of lasting 'conservation covenants' between the two main parties? To address this question, I describe a few scenes from this play with which I happen to be familiar as a result of my own participation in the conservation policy process over the course of the past decade. From my reflection on these snatches of dialogue, I shall then try to extract some messages that might serve to advance the terms of the debate between anthropologists who have an interest in conservation and conservationists who have an interest in anthropology.

The road to hell is paved with logs

The latest phase in PNG's conservation policy process began in April 1990, when the World Bank orchestrated a round table meeting that gave birth to a creature that was eventually called the National Forestry and Conservation Action Programme. One of the few unexpected outcomes of that meeting was a resolution to establish a task force on Environmental Planning in Priority Forest Areas, with a mandate to take immediate action to counter the threat of logging in areas which were thought to contain plentiful amounts of biodiversity. The Task Force spent most of its initial donor funding on a couple of expeditions to potential conservation areas in Milne Bay and New Ireland provinces, and in each case, an anthropologist was sent along to reflect on the feasibility of achieving 'formal landowner acceptance' of a conservation strategy which had not yet been devised (Filer 1991b; Young 1991).

And so it was that three national stakeholders and I descended (literally, by helicopter) on the people of the Lak census division, at the southern extremity of New Ireland, in August 1990. Our first port of call was the home of the locally elected member of the New Ireland provincial government, and the first of our three public meetings was held that night beneath his house, as the rain poured down around it. Our host warned us in advance that his own personal interest in protecting the local environment was not shared by most of his constituents, and that his own survival as an elected politician depended on his ability to represent the views of those people who 'pull the strings on the backs of us leaders'. If he was referring to the village elders who later joined the discussion, then we could understand why he had been persuaded to facilitate the deal between a Malaysian logging company and the local landowner company that was already awaiting the imminent approval of the relevant national government ministers. These old men could only talk of how much they had 'suffered' from the experience of living in the 'last' (most backward) corner of the province, if not the whole country, of their immediate need to secure some small amount of money to ease that suffering before they died, and of their absolute determination to sell their trees in order to get it. Their brains were pained by any thought of an alternative.

And what was the alternative? Not having a precise answer to this question, my colleagues and I did our best to improvise a floor show that contained a number of compelling images without making any specific promises. Talk about laws, policies, permits and agreements would surely lower the local brain pain threshold to an unacceptable degree. Our time was short, and so was the patience of our audience. So we talked about roads instead, not only because the logging company was promising to build them, but also because the Tok Pisin word *rot* provides the metaphorical key to almost every thought about development in rural Melanesia. And we talked about the Bible as well, because this was the most familiar source of parables and homilies on such topics as poverty, greed, and deliverance, and hence, perhaps, on more elusive concepts such as 'conservation' and 'sustainable development'.

In our exposition of the two-road theory of development, those who travel the logging road, which looks so wide and straight at first, will sooner or later find that their progress is cut short by fallen trees, and that the road itself does not last very long, as logging roads and logging royalties both run out very fast in PNG. So this is the road to hell, or maybe a road that goes round in circles, and therefore does not get you out of the last corner. The better road is the long and winding one, where the long-term benefits are greater, but so are the short-term sacrifices. The last people to make the choice could be the first people to choose the right road, and the first to discover the delights of the 'benefit package' with which the international community (or maybe God) was preparing to reward them.

In all three public meetings, we found that this kind of semi-Biblical talk had greater resonance than some of our other spontaneous experiments in landowner awareness. For example, our audience saw little merit in the suggestion that future generations of local landowners might regard the decision to sell a lot of trees to an Asian logging company in the same way that the present generation regarded the earlier sale of land to European colonizers for the price of an axe or a piece of cloth. On the other hand, our talk about 'another road' provoked their recollection of the many empty promises which the government had previously made, and our own association with the national Department of Environment and Conservation might well be taken as evidence that we were just the latest in a long line of hot air merchants.

One member of our team tried to overcome this obstacle by talking about a rare species of butterfly, which, according to his estimate, could be captured and sold for 50 kina (then worth more than 50 US dollars) a piece on the world market. In order to reinforce the point, he captured one himself and carried it from one meeting to another, using it as an illustration of the potential economic value of an unlogged forest. But that butterfly was the only visible and concrete thing we had to offer. The rest was just talk, and might mean nothing, as members of our audience constantly reminded us. We argued that the logging company had also made many promises that might prove to be false, and that the loggers had more reason to deceive the people than we did, because they stood to make a profit out of their promises, whereas we did not.

But the loggers had already provided benefits, including quantities of money, which had much greater value than the sight of a butterfly and a lot of talk about another road.

The Task Force decamped from Lak with a vague promise to return with a map which would represent the compromise proposed by the provincial member at our first wet meeting – that a way might be found of dividing his electorate into areas which could be logged and areas in which the 'other road' could be pursued. But my own faith in the feasibility of such a deal had been thoroughly shaken. Shortly afterwards, I wrote a newspaper article in which I made the following observation:

> Our conservationist friends in the developed world sometimes seem to imagine that Papua New Guinean villagers resemble the Indians of the Amazon rainforest – people living in simple harmony with Mother Nature, for whom 'development' is a menace imposed by outsiders. If these people went to Lak, they would be in for a big shock. It would probably be difficult to find another place in PNG where local landowners were more insistent on the need to have their trees cut down as soon as possible!
>
> (Filer 1991a)

By the time that this article was published, in January 1991, the Minister for Forests had already issued a timber permit to the local landowner company, and the cause appeared to be lost. The Task Force never did return to Lak, nor did it ever go anywhere else, so the scepticism of our local audience would seem to have been justified.

But this was not the end of the local story. In August 1992, long after the task force had been disbanded, the provincial politician sent a letter to the International Tropical Timber Organization, asking for help to control the environmental damage caused by the logging operation which was now under way, and to establish a 'Conservation Foundation' and a 'Wildlife Management Programme' incorporating profitable environmental activities'. This letter eventually found its way back to the desk of a New Zealander who had just been appointed as Chief Technical Adviser to a new Biodiversity Conservation and Resource Management Programme which was to be housed in a Conservation Resource Centre attached to the national Department of Environment and Conservation. This entity had just been granted five million US dollars by the Global Environment Facility, primarily for the purpose of establishing a pair of experimental 'integrated conservation and development projects'. The Western conservationists associated with the programme read the member's letter as 'a social invitation from a unified group of landowners eager to explore alternative methods of forest development' (McCallum and Sekhran 1997: 19). So they spent the next three years in what eventually proved to be a losing battle with the loggers and their local allies. This seems somewhat strange, in retrospect, because the Chief Technical Adviser and his national counterparts already had copies of my own detailed report on the activities of the Task Force (Filer 1991b), which certainly did not support their chosen reading of the member's letter.

> One can only conclude that their decision to return to Lak was either the result of bureaucratic inertia and structural amnesia, or else a deliberate move to 'take the fight to the enemy' by taking the local politician's words at face value, and seizing the opportunity to mobilise a re-enlightened community against the forces of darkness which it had previously been so eager to embrace.
>
> (Filer and Sekhran 1998: 248–249)

At any rate, the lessons learnt from their first big experiment (McCallum and Sekhran 1997) confirmed at least three of the lessons learnt in the brief encounters of the Task Force: the best (if not the only) way to persuade landowners that conservation is a good thing is to represent it as an alternative form of development; it is very difficult to represent conservation as an alternative form of development without raising expectations or making promises about the delivery of such development; and it can also be very difficult to engage landowners in a dialogue about conservation and development options without exaggerating the intensity of factional struggles within their communities.

The excavation of rural desire

The steadily unfolding failure of the Lak experiment did at least cause its conductors to take more interest in the stories told by anthropologists about the people who live with lots of biodiversity in their backyards. When the Conservation Resource Centre decided to initiate its second project in Madang Province, on the northeastern slopes of the Bismarck Range, I was invited to join the first patrol to the 'area of interest' in April 1995 (Filer *et al.* 1995), and then to assist in the design of a social feasibility study which was to form an integral part of the 'Bismarck-Ramu project'. By the end of 1995, the lessons of Lak had been absorbed into a framework plan for the new project, which asserted that:

> ICAD community liaison staff must start by not talking about conservation, not talking about economic incentives, and not talking about development. . . . Instead they should be listening to the community, and facilitating a debate within the community as it explores its own needs and beliefs . . . Conservation and the establishment of protected areas will be the outcomes, not the motivators.
>
> (quoted in Ellis 1997: 16)

If conservationists could not hope to outbid developers for access to natural resources on customary land, they should stop thinking about conservation as an opportunity cost for which landowners ought to be compensated, and start thinking instead about the pursuit of something called 'genuine community conviction'. And in the Bismarck-Ramu case, it looked as if they had the time and space to test the weight of such moral incentives, without having to counter the material incentives offered by immoral developers.

This strategic innovation coincided with the appointment of a new Chief Technical Adviser to the Biodiversity Conservation and Resource Management Programme. Unlike his predecessor, whose international career had been launched on the back of his experience as a park ranger in New Zealand, the new man was an applied anthropologist whose main qualification was his experience of working with an Aboriginal Land Council. Unlike his predecessor, he also knew a good deal about things like Participatory Rural Appraisal (PRA), and might even have shared the responsibility for insisting that all the stakeholders now mixed up in the Bismarck-Ramu project, including himself, should participate in a workshop on this subject in order to refine a new form of 'community entry'. The PRA Toolkit Workshop was duly held in May 1996 (Grant 1996), and the Bismarck-Ramu project began its new lease of life as an exercise in the discovery of self-reliance.

The workshop served as an initiation ceremony for the new members of the Community Development Team that was to test the level of community conviction in the designated Area of Interest. The second stage of their initiation was a training programme orchestrated by an American Community Development Trainer who had a long track record of awareness work in PNG village communities. His own written records of this training exercise feature the deconstruction of two texts produced by other Western conservationists associated with the project. The first of these was an essay written by an American entomologist who now doubled as the project's education coordinator. This essay (Orsak 1997) took the form an imaginary dialogue between conservationists and landowners, in which the latter presented a familiar litany of reasons for not listening to the former, and the author suggested a number of ways in which the conservationists might rephrase their arguments in order to avoid such reactions. One of the points made in this essay was that traditional Melanesian culture lacks the sort of 'conservation ethic' that might encourage landowners to accept the arguments of conservationists in a modern political setting.

But members of the Community Development Team were reportedly unable to understand, or unwilling to accept, this argument. They felt that Melanesians were accustomed to think of themselves as being part of their natural environment, rather than being separated from it, and this meant that there could be no point in making a distinction between the idea of people looking after their forest and the idea of 'bush spirits' (*masalai*) looking after it, or even the idea of the forest looking after itself (Lalley 1998). The Community Development Trainer took this to be a sign of the gap which really exists between Western and indigenous conceptions of the natural environment, rather than a function of the role which the trainees were now expected to play as mediators in the conversation between conservationists and landowners. The trainees themselves were disposed to expect that indigenous knowledge and community conviction would be found together in the bed of self-reliance, once they got beneath the mental blankets of 'development-dependency'.

The second text whose deconstruction helped to shape their vision was the opening speech that project staff had made at village meetings during previous patrols to the Area of Interest. The result of this exercise was a prototype or paraphrase for the opening speeches that would henceforth be made by members of the Community Development Team:

> We are not working for a bank, a logging company, a mining company, a church, or a political party. We are not even working on a 'project', we are simply an 'environmental group' (*environmen lain*) working with the Department of Environment and Conservation. We are not going to help you to 'conserve' or 'develop' your 'resources', nor is the 'government' going to do so. We are only here to talk. We don't know what you know. Perhaps we can teach each other something. Perhaps we can help you to help yourselves. And we will only stay here if you want us to.
>
> (CRI 1996a, Appendix 3)

This display of verbal diffidence was intended to prevent the listeners from engaging in what the team members later came to describe as a process of 'jumping through hoops' in order to capture the windfall benefits which villagers have learnt to associate with passing projects. But it also contained an element of bad faith in its own right, because it concealed the ultimate objective of a donor-funded conservation project, as if this were a secret mystery which could only be revealed to those villagers who learnt to jump through another set of hoops in order to demonstrate their 'conservation conviction'.

The Community Development Patrols were sent out at intervals of approximately two months. In each case, the Community Development Team spent a few days in a joint 'briefing session' under the supervision of the Community Development Trainer, and was then divided into two or three smaller groups, each spent about three weeks visiting a number of local communities in the area of interest, before finally returning to base for a 'debriefing session' in which their findings were recorded for posterity. The 'community development process' was itself divided into seven steps, each of which required a separate visit to any given community (Ellis 1997: 21). During the first two visits, the interaction between team members and local villagers was conceived as a process of 'story-telling', 'trust-building', and the 'hearing of community themes' (ibid: 30). The more elaborate instruments in the PRA Toolkit were saved for the 'diggings' undertaken in subsequent visits, lest they be construed as 'gimmicks' by an unfamiliar audience (van Helden n.d.).

After eight rotations of the fieldwork cycle, 34 communities had received at least one visit from the Community Development Team, but very few had reached Step Five, which was the point at which they were able to formulate an 'Action Plan' to address their own problems in a self-reliant manner, and hardly any had ascended to Step Seven, at the very top of the ladder, where the role of project staff would supposedly be limited to monitoring and facilitating the implementation of this plan. That is partly because most of the communities had received less than five visits by the end of 1997, but in some cases, the process had been cut short by the failure or refusal of local villagers to play the game according to the rules developed in the briefing and debriefing sessions which took place backstage.

Backstage was the place where nearly all the Western conservationists associated with the project stayed throughout this period, in order to avoid being cast in the role of 'bosses' whose white skins might be taken to presage the delivery of 'cargo' (CRI 1997b: 6). The only significant exception to this rule was the Social Feasibility Consultant, a Dutch sociologist attached to my own division of the National Research Institute, and even he was not allowed to enter a community until its members had safely climbed the first two rungs on the ladder of self-reliance. His role was to document the findings of the barefoot anthropologists in the Community Development Team, to verify or supplement these findings through his follow-up patrols, and then present the outcome as a descriptive account of social relations in the Area of Interest (van Helden 1998a) and a more prescriptive assessment of the 'social feasibility' of establishing a conservation area within it (van Helden 1998b).

Two fundamental 'community themes' were dug up from this field of inquiry. One was the relationship between thousands of 'Jimi people', including those resident in the upper Jimi Valley and those living on the margins of the Ramu floodplain, and a few score 'Ramu people', most of whom were living along the banks of the Ramu River. At one stage, members of the Community Development Team described this relationship as an instance of 'neo-colonialism', in which the Jimi people were colonizing the Ramu people in both a physical and cultural sense, as most of the 'original landowners' had already begun to speak the language of the newcomers who were taking their land and plundering their resources, and the highlander's fear of lowland sorcery had become their last line of defence (CRI 1996b: 45). The second theme was the intense political rivalry, and occasionally violent conflict, between the clans and subclans into which the Jimi people were divided at all altitudes. For those who formed the colonizing frontline in the Ramu floodplain, these political relationships were further complicated by the fact that a minority did have a solid claim to be the traditional owners and occupants of the Bismarck foothills, and even segments of the floodplain itself, while the majority did not (van Helden 1998a: 236–239). In this social and political context, the Community Development Team made the somewhat unremarkable discovery that some of the Ramu people had an interest in the conservation of indigenous knowledge and natural resources, while most of the Jimi people were primarily concerned with the distribution of wealth, status and power.

In retrospect, the weight of demands with which the Jimi people ultimately crushed the community development process might be blamed on the expectations which had already been raised before that process was initiated, and might thus be taken to confirm the principles by which it was designed. In the Jimi Valley, the Community Development Team was almost ambushed by a gang of youths who thought it was a 'group of company officials or agents' bearing large quantities of money (CRI 1996b: 27–28), and was later confronted with demands for compensation for the 'theft' of natural resources by the 'foreigners' who had conducted a biological survey in October 1995 (CRI 1996c: 37). And on the Ramu floodplain, the team was constantly embroiled in a continuation of the squabbles about payment and employment that had first erupted when the project base was set up next to the half-built airstrip back in February 1996 (CRI 1996b: 30–31; Ellis 1997: 29). But it soon transpired that the root cause of such misunderstandings and disputes could be traced back to the employment of the airstrip's local architect, a university graduate, as the project's Community Facilitator, and their subsequent escalation was due to his own determination to 'capture' the project in order to win votes at the national election which was due to be held in June 1997.

The Community Facilitator was duly summoned to headquarters and placed behind a desk, and therefore had to quit the project to continue his electoral campaign. But his method of advertising the project's existence still served to undermine its objectives. His claim to be the 'boss' who had secured a huge amount of project funding from the World Bank (CRI 1997b: 16) caused all sorts of rumours to invade and obstruct the Jimi people's participation in the community development process. There were some who partly believed his claim, and wanted to know why they were not getting a bigger share of this 'free money'; there were others who were somewhat sceptical, and wanted to know when they would get to meet the 'real boss'; and for those who did not trust this candidate at all, there was the option of believing that what he had really done was to help the government sell their land to the World Bank (CRI 1997a: 10, 15, 42). Those members of the Community Development Team who hit this wall of ideas in the Jimi Valley were also warned that their lives were in danger, so they beat a hasty retreat.

The concept of the project did not enable the excommunicated community facilitator to win the election. But even after the votes had been counted and the rumours had subsided, the Community Development Team was still lamenting its inability to wean the Jimi people away from their 'handout mentality'.

> The team's observation of the attitudes and body language of people during meetings gave an indication that there is really little interest in any community development if it requires the people to get organised or do actual work. The community has said again that, 'The CDT come and come again and nothing is given to the villagers'. It appears that in spite of

> the amount of time put into the village development process, the communities still have an overwhelming reliance on 'cargo' and they are waiting for the CDT to bring something.
>
> (CRI 1997b: 39)

The persistence of such 'cargo expectations' was not only linked to the many promises which had just been made by candidates competing in the national election, but also to the promise of drought relief which followed shortly afterwards, to the experience of an influenza epidemic which also afflicted the area in 1997, and to the more abiding shortage of both natural resources and development opportunities in the Jimi Valley (Ellis 1997: 43; van Helden n.d.). But when one reads the reports of the Community Development Patrols, one finds that the number of conversations in which the Jimi people wanted to know what the team had to offer is matched by the number of conversations in which they wanted to know what the team really wanted. For example:

> They always want to know exactly what we are after. Their never-ending [enquiries] make us feel like we have confused them more than ever by coming in and working with them ... [They] keep coming and asking questions such as: 'After you had visited us for so many times and have had enough of that, what will come next? What are you going to do; what are you going to establish? ... '.
>
> (CRI 1996b: 41)

And it sounds as if they never could get a straight answer to such questions.

The talk presented by the Ramu people was considerably less demanding, but in some ways more mysterious, less political, and more religious in its flavour. While most of them had not been 'spoilt' by the sight and sound of Western conservationists during the previous phase of the project, the concept of the 'cargo cult' was no less evident in their reactions to the process of 'community development'. But while the Community Development Team used this concept to explain the Jimi people's failure to realize or practise the virtues of self-reliance, the Ramu people used it to explain, and in one case to reject, the process of discovering these values. In other words, the Ramu people thought the project which was not a 'project' was perhaps a cult, whereas the Jimi people knew the project was a 'project', or subscribed to the belief that it must be a 'company' of some sort, and demanded that it yield the sort of benefits which members of the team called 'cargo'.

When the Community Development Team made its second visit to one Ramu river hamlet, the headman was quite effusive in his greeting:

> My dear children, I am greatly delighted that you had cared enough to visit me again as you had promised me the last time. What better time could you have chosen to visit me and your brothers and sisters here, than during

> such a time like this when I am no better than dead.... You must NOT leave us, never, remember that we are family now. We are on the ground now; we haven't yet set foot on the first step on the ladder leading up to that house you mentioned. And the door into the house like you mentioned is way up there out of our reach. Just think of the taro. When you plant it, does it grow, mature, bear and ripen overnight? Not that I know of! Or do you know of such a 'miracle' taro? My children, let's talk reality and stop dreaming! In the same way as we wait for the taro we plant to grow, mature, bear and ripen; you and I (us) must work hard and take our time. We must NOT rush, for rushing is risky, as far as my experience makes me believe.
>
> (CRI 1996b: 42)

One of the headman's sons glossed this welcome by declaring that the team had been brought to them by Jesus Christ himself, and made allusions to the parable of the sower and the seed:

> You are the sower and your words are the seed. We who listen, for our part, are the different types of soil. What we do after we hear your words and even after you had gone will show what kind of soil we each are. If we are good soil then the seeds you sow will grow and bear fruit.
>
> (ibid)

Encouraged by such thoughts, the team set off to make its initial visit to a neighbouring hamlet, whose inhabitants were also 'surprisingly hospitable' and exceedingly devout. On the other hand, their spokesman's response to the team's explanation of its 'work' was quite at odds with their initial hospitality:

> My people and I have the right to know whoever comes into our land and also why he or she comes in. We asked you to clarify to us your work and you claim that you have done so. The fact is, we are even more confused, because your words are like complete nonsense, they are absolutely meaningless.... How can we possibly make our decision/stand if we are not clear about your work? Therefore we ask you not to come again here in the future ...
>
> (ibid: 44–45)

So the team members said 'thank you very much' and quit the scene.

They were understandably puzzled by this negative outcome, which was apparently unique in their experience of first encounters. But a partial explanation emerged during a subsequent patrol along the banks of the Ramu River, when the team members got a 'very cool reception' in two other hamlets, including the one whose headman had previously spoken of them as his children. Now the headman addressed them as 'spirits of the dead people' (CRI

1997a: 21), and they were given to understand that the community which rejected their initial advances had since

> spread rumours about the CDT's being members of a cargo cult known as 'Bembe'. The suggestions were that we were spirits that had come back from the dead (*daiman*) and sinners (*sinman*) and that people would fall ill if they spent the money that we had paid them for accommodation. People were obviously very confused because last time the CDT's had come and when somebody had been ill, they had given them medicine and had treated him well. People initially did not know whom to believe and waited without giving us food until they had heard our side of the story. It seems that our exercises, including going back and forth to town and resource mapping had backfired on us in the sense that local people had interpreted them as part of the magic that we were using to bring them under the spell.
>
> (ibid: 42)

But the headman himself, who seems to have believed this story, still wanted the team to make regular visits, 'otherwise the communities would lose interest in us' (ibid: 21). The team members thought it was 'quite ironic that the strong emphasis on avoiding unrealistic expectations and self-reliance makes people think we are a traditional cargo-cult', but they also found something more rational and sinister in the hostility of one local leader who was 'strongly in favour of logging and had rightly identified us as potential opponents to such activities' (ibid: 42–43).

Whatever their real motivation, those Ramu people who espoused, or were said to espouse, the ironic perception of the project as a sort of cargo cult did not reach the top of the ladder which led to the 'house of conviction'. The first signs of this achievement were detected amongst the small group of Ramu people encountered by the very first patrol to the Area of Interest, living in immediate proximity to a large bunch of very demanding Jimi people. And for this very reason, their convictions had a question mark against them, because the community workers and project managers could see them 'not as a drive for conservation alone, but as a way of addressing the major social problem of integrating the Jimi settlers, with their aggressive land use patterns, into the Ramu way of life' (Ellis 1997: 49–50). The minutes of meetings held at the end of 1997 suggest that it might not have been a 'drive for conservation' at all. When the Community Development Team met with members of the two landowning clans, and told them that 'the Project would only be involved with them if they want to conserve their natural resources', the landowners evaded this question by asking that 'the team should not meet and work with the Jimi settlers' (ibid: 56). The team disregarded this request, and held a separate meeting with the settlers, in which they rehearsed the substance of their earlier discussion with the landowners.

> The settlers agreed and supported the landowners' idea of conservation. They reported that there were rumours that some of the landowners had interests in logging. . . . The settlers see that if logging is brought in they will be adversely affected. Each family has a very small area to spare for logging. Secondly, the areas that they normally hunt in which are beyond what they were allocated by the landowners might be destroyed by logging. They will find it very hard to continue living the way they do now. Hunting and gardening practices would be affected.
>
> (quoted in Ellis 1997: 56)

When the Conservation Area Manager finally came to talk to the landowners in April 1998, they had already reconstructed their desire for conservation as a desire to make provision for the livelihood of 'future generations', thus neatly aligning themselves with the Fourth Goal of the National Constitution, but this was still not enough to convince the conservationists of their sincerity (ibid).

Greater progress seems to have been made in meetings with the residents of another village on the Ramu River, even though their interest in conservation also stemmed in part, if not entirely, from their wish to keep or get the Jimi people off their land. This was the village where people could clearly remember the original visit by an officer of the Australian colonial administration

> who discussed the establishment of a National Park and told people to limit their hunting in certain areas and not to use guns. The Ramu river communities still purport to observe these rules but are very concerned about the continuous hunting on their lands by Jimi Settlers. It is not entirely clear whether they understand the full concept of a national park or whether their interest in such is mainly a result of the need to stop the Jimis from further encroaching on their ancestral lands.
>
> (quoted in Ellis 1997: 51)

These people had reiterated their interest in the establishment of a 'national park' when a team of government officials came back to assess the original proposal in 1986 (Filer *et al.* 1995: 47), and repeatedly brought the matter up in conversations with the Community Development Team. Indeed, one might well wonder what difference, if any, the team actually made to these people's stated preference for 'conservation', apart from persuading them that the government would not do very much, if anything, to help them realize their goal. It is true that they had to prove their own capacity for self-reliance by meeting the cost of a trip to Madang to meet the Conservation Area Manager, and when this action was rewarded by his return trip to the village, the villagers had found some more 'convincing' reasons for their preference. Apart from defending their resources against the encroachment of the Jimi people, they also wanted to reinforce clan boundaries within their own community, they were worried about the possible impact of a large-scale mining project, and they saw some potential for attracting tourists to the area. But when the Conservation Area Manager raised the question of whether traditional beliefs and practices would provide a suitable basis for conservation, one villager replied as follows:

> We cannot use the same practices used by our ancestors nowadays, because educated people do not respect and listen to the village elders. Educated people are proud of themselves, they think that they have been to school and are more knowledgeable than the village elders. Therefore if village elders make rules to conserve a certain area, people that have some form of western education will not adhere to those rules.
>
> (quoted in Ellis 1997: 57)

This sentiment did not exactly fit the bill to which the Community Development Team was wedded at the outset, which was to find or plant 'community conviction' in the rediscovery and reproduction of indigenous knowledge.

At least the Ramu people showed a lot of interest in the subject of indigenous knowledge, even if they were sometimes inclined to reject it as the work of the devil. But the manner in which the subject was broached in their conversations with the Community Development Team would only have confirmed the Jimi people's prejudice against them. On one occasion, as they set off to visit a Ramu community, the team members received the following piece of advice from their Jimi hosts:

> They will not feed you, they will not house you. And if they ever do house you, this will mean you sleeping underneath their houses. They don't even have food anyway, for they know no gardening. We were the ones who taught them gardening and all other skills; we civilized and 'tamed' them. They live only on sago. They can't even talk sense with you like we do, and they will never understand your message.
>
> (CRI 1996b: 41)

What made 'sense' to Jimi people, by and large, was talk about 'development', which members of the team refused to talk about because they thought that it was just another name for 'cargo' (Ellis 1997: 36). And what did not make sense to Jimi people, by and large, was the team's inclination to talk about indigenous knowledge as knowledge about 'the bush', about nature rather than culture, and to talk about its value in spiritual, rather than purely practical, terms.

Yet I would not go so far as to say, from the evidence available, that the Jimi people had their feet firmly planted on the ground, while the Ramu people had their heads in the clouds or their minds in 'the bush'. On both sides of the fence, villagers were equally concerned to search out ways of establishing material exchange relationships with the Community Development Team, and seem to have been equally puzzled by the team's reluctance to reciprocate (CRI

1997a: 11; van Helden 1998a: 258). And it would be hard to argue that one side had more or less of an interest in the prospect of large-scale resource development, whether by logging, mining or petroleum companies. In this respect, the difference resided in the fact that such companies really did show rather more interest in the resources claimed by the Ramu people than in those claimed by the Jimi people, and if the latter really did believe that an Iraqi logging company was about to harvest the timber in the upper Jimi Valley (CRI 1997b: 15), then theirs was by far the greater delusion.

The bottom line, if such it can be called, is that the people who were most convinced about the need for conservation were Ramu people who harboured a desire to roll back the tide of territorial encroachment by groups of colonizing highlanders, and were much less concerned about the threat posed by foreign resource developers (van Helden n.d.). Proponents of the project's community development strategy would say that it does not matter why people want to conserve their resources, so long as they really want to do it (Ellis 1997: 54). But then we have to ask whether this conviction stems from the correct application of a PRA Toolkit, and whether those who are convinced have also got the power to achieve their goal when they have learnt that no one else is going to help them if they cannot help themselves. However long the Community Development Team spent 'digging' into the local social soil with its 'timelines' and 'resource maps', what it found at the bottom of the hole was something already present on the surface – a set of disputed territorial boundaries and unequal spatial relationships.

Selling the biodiversity business

The Conservation Resource Centre was in fact a rather peculiar creature in the conservation policy process precisely because it was housed in the national Department of Environment and Conservation, and therefore had the benefits and costs of being seen as part of the 'government' that normally failed to meet the expectations of rural villagers. Most of the other ICDPs in PNG have been designed and implemented through partnerships between national and international NGOs, whose own specific aims and interests have variously coloured the nature of their dealings with local landowners. Three of these projects received substantial financial support from an organization known as the Biodiversity Conservation Network (BCN), whose approach to the business of establishing conservation areas in PNG was markedly different to that which had recently been espoused by the Bismarck-Ramu project. These contrasting strategies or scripts alert us to a range of possibilities for local variation in the substance and outcome of the talk which has been going on between local landowners and Western conservationists.

The BCN was initially established as a five-year programme in late 1992, but the period of implementation was later extended to six and a half years, which meant that the programme came to an end in the first half of 1999. Its funding was actually treated as part of the US government's 'attribution' to the Global

Environment Facility (BCN 1997: 120), but it was implemented as one component of the more enduring Biodiversity Support Program, which is sponsored by a consortium of three American NGOs – the World Wildlife Fund, The Nature Conservancy and the World Resources Institute. During the six and a half years of its existence, the BCN provided implementation grants to a total of 20 conservation projects in the Asia-Pacific region, in order to evaluate the effectiveness of what were described as 'enterprise-oriented approaches to community-based conservation of biodiversity'. This meant testing a 'core hypothesis' that stated that 'if local communities receive sufficient benefits from an enterprise that depends on biodiversity, then they will act to counter internal and external threats to that biodiversity' (ibid: 1).

The various NGOs that were on the receiving end of these grants were apparently required to volunteer some of their staff for a form of initiation into the mysteries of the project cycle (that ceremonial practice which is familiar to denizens of the aid industry) in order to improve their capacity for 'adaptive management'. The initiates were thus taught to rethink their projects (and now also think about how to spend the BCN grants) by following the path that leads from the production of a mission statement to the process of revising management and monitoring plans in light of lessons learnt from their implementation (Margoluis and Salafsky 1998). The second stage of this initiation involved the production of a 'conceptual model', which consisted of a lot of boxes containing the various factors which had some positive or negative impact on the conservation of biodiversity in the place where a given project was located, and a smaller number of circles containing the specific project 'interventions' which were intended to modify the causal connections between these factors, and thus to test the core hypothesis.

In January 1999, as the BCN programme drew to a close, Network personnel convened a public meeting to reveal the lessons which had been learnt from their experiments with the core hypothesis, and to assess the extent to which various threats to biodiversity had actually been diminished through the pursuit of each project's enterprise component. After the newly anointed Melanesian Masters of Business Administration had recounted their own experiences in the biodiversity business, their patrons in the Network offered the audience a number of 'general but non-trivial guiding principles', which took the form of 'conditional probability statements':

> For example, we might say 'In Melanesian type social systems, it is generally better to work with the big man to solve conflicts unless he is corrupt'. The key features here are that the principle applies to more than one place (in Melanesia) but not everywhere. Furthermore, it is not guaranteed to work in all instances – the user has to be smart enough to apply it to his or her own situation – for example, to determine if the big man is corrupt or not.
>
> (BCN 1999: 6)

Two particular points struck me as I listened to this presentation. One was the evangelical zeal of the American conservationists associated with the Network, which had presumably permeated all their previous conversations with the Melanesian conservationists whom they had been initiating into the business of turning biodiversity into a commodity. The other was the fact that local villagers were represented, by the Melanesian conservationists, as people who still had a great deal to learn about this new form of entrepreneurial culture, and who still had to be persuaded to abandon a variety of economic practices which were themselves regarded as major 'threats to biodiversity'. If the Melanesian conservationists had been preaching with the same vigour as their American mentors, and thus breaching one of the main taboos imposed on the Bismarck-Ramu project, they had not yet won too many converts from the communities that hosted their projects.

My only contribution to the public action in this scene was to ask one of the national staff employed on the Lakekamu Basin project why he thought that small-scale alluvial gold mining was such a big threat to biodiversity, and how he expected the local villagers, who had obtained the bulk of their cash income from this activity for many decades, to sacrifice it in exchange for money to be earned from ecotourists who had not yet put in an appearance. His answer to the first question was that people who practice small-scale alluvial mining are more likely to welcome the prospect of a large-scale industrial mining project than people who do not. His answer to the second question was that it was very difficult.

The small miners in question had originally learnt their craft from the European prospectors who were active in this area during the second and third decades of the twentieth century, and who then vanished from the scene (Nelson 1976). So their knowledge of mining was not exactly 'indigenous', even if it was by now customary, and since 'we all know' that mining is a dirty business, it was evidently not the sort of knowledge which had any sort of place in an 'entrepreneurial culture of conservation'. But 'we all know' that logging is a dirty business too, especially when it is done in 'virgin tropical rainforests'. So if Western conservationists feel the need to wean indigenous miners off their gold pans in order to save them from opening their arms to large-scale mining companies, how can they also be selling the virtues of walkabout sawmills to villagers who might otherwise open their arms to large-scale logging companies? For that is what they do in many other Melanesian conservation projects, where knowledge of small-scale saw milling is not even traditional, let alone indigenous.

No one at the meeting had a ready answer to this final question. Perhaps the answer lies in the physical or conceptual difference between rocks and trees. Eco-forestry adds economic value to the forests which contain biodiversity values, and if we get our definitions right, it simultaneously causes its practitioners to appreciate the value of the biodiversity in the forests which they are only very gradually cutting down. Mineral resources have no place in such equations, and the pursuit of gold, by any means at all, is nothing but a testament to human folly.

The gold-panners of the Lakekamu Basin had not yet had much opportunity to evaluate this line of reasoning, because the new business on offer was ecotourism, not ecoforestry. Or perhaps it was no new business at all. As one of the national conservation project staff described the prospect:

> After the first year, we had a training session about ecotourism. One of the guys got all inspired and wanted to build a guesthouse. He organized his family and built the place. This was really hard. I was glad that he was showing interest, but I was worried about not having any guests come. They built the house and then they started asking when the tourists would come. I didn't know what to tell them. And this has turned out to be a problem. Only a few people have come and already the guesthouse is starting to fall apart. I feel responsible for what has happened and that I let them down. Even now, when I go back to the Basin, they ask me, 'When will they come? Is there any news of tourists coming?'
>
> (BCN 1999: 166–167)

While this family waited for Godot, the alluvial miners, crocodile hunters and betelnut farmers went on minding their own business, mindful perhaps of other promises and prospects of 'development' which had previously come to nought.

Back in 1988, I conducted a social impact assessment of a medium-scale industrial mining project in this area, in which I made the observation that

> it is better to promote those money-raising activities which have already proved viable, and even those which have previously failed for identifiable and remediable reasons, than to promote novel activities whose viability is an unknown factor.
>
> (Filer and Iamo 1989: 58)

The project in question never got past the design stage, but I certainly did not find that this prospect of 'development' held any special appeal for the alluvial miners in the area, who had tried and failed to obtain the mining company's assistance in getting their own product to market, and who were understandably concerned that it might later find some way to put them out of business. Little did they know that, when the mining company had gone, a conservation company would come and look for ways to do so.

Disintegrating conservation and development

I have chosen to represent the conservation policy process as a play full of odd characters and strange talk, not because I wish to emphasize the absurdity or futility of that process, but because I want to underline its unpredictable, disputed, open-ended quality. Some anthropologists might prefer to deconstruct the narratives contained in that tedious mass of project and policy documents which is produced by global actors like the Global Environment Facility (Zerner 1996). But this kind of discourse analysis only seems to reinforce the semblance of imperial power that it purports to criticize (Grillo and Stirrat 1997). The discourse of 'integrated conservation and development' actually disintegrates in talks between the 'stakeholders' who do not live in Washington and do not read academic journals. It is only by participating in these conversations, both as listeners and speakers, that we can see how conservation policy is manufactured on the national and local stages where the global script does not dictate the outcome of the play (Croll and Parkin 1992: 33–34).

Many anthropologists have made it their business to debunk what Ellen (1993: 126) calls the 'myth of primitive environmental wisdom', and Melanesian specialists (myself included) have been amongst the most vociferous exponents of this stock in trade (Bulmer 1982; Dwyer 1982, 1997; Allen 1988; Filer 1991b, 1994). But what concerns me here is not so much the truth as the politics of this sceptical approach to the value of indigenous knowledge in the conservation business. Brosius (1999) has recently suggested that the deconstruction of essentialized images serves no useful purpose if it only demonstrates our own exclusion from the policy arena, and might, in certain circumstances, offer extra ammunition to the forces of darkness. Instead of quibbling while Rome burns, we need to ask how anthropologists can help to fight the fire.

The Melanesian version of Rome does not resemble the Malaysian political setting in which Brosius has situated his own argument, even if it does contain Malaysian logging companies. Anthropologists and conservationists alike have far more freedom to criticize the loggers and the government, and there is much less chance that anything they do or say will jeopardize the rights or compromise the strategies of customary landowners. Where Western anthropologists play many active roles in the conservation business, they may not always see the value of indigenous knowledge in the same light. Yet their views, taken as a whole, still tend to diverge from those espoused by the national conservationists employed in the same industry, who are more inclined to regard the 'myth of primitive environmental wisdom' as a rock on which to build effective rhetoric and policy.

The national conservationists have good cause to take this position, if they believe their point of view will maximize the flow of foreign funds that keeps them in employment. Occidental donors have no cause to challenge myths that help to build the confidence of their allies in the national policy community. This is one reason why the donors are prepared to pay for large numbers of national 'community workers' to be trained in the mystical arts of PRA, displacing the more rapid and sceptical appraisals offered by alien anthropologists. But the more time that is spent gathering local participation in a positive appraisal of indigenous knowledge, the greater the risk of seeming to patronize local landowners who think they need to know something more than what they think they know already (Filer and Sekhran 1998: 331). The subjects of this process have every right to wonder why outsiders of any complexion are being paid good money for such work, when it seems to demonstrate the limits of their own access to the donor dollar.

At the same time, the concept of indigenous knowledge has gained currency in the Melanesian conservation business in a way that tends to make the adjective prescribe or circumscribe the object of the knowledge. It has to be knowledge of what and where the wild things are, because that is the knowledge that matters to conservationists who want to keep those wild things as they are. It cannot be knowledge of subsistence farming or alluvial mining, let alone knowledge of 'cargo cults' or electoral politics, because such knowledge cannot be used for this purpose. But if we cease to think of Melanesian landowners as 'rainforest peoples', and see them as they generally see themselves, as 'gardening' or 'farming' peoples, then we are obliged to recognize that the productive powers of their indigenous or traditional knowledge are weighted heavily towards the art of cultivation, and that the process of 'development' has only added to this bias.

No wonder, then, that when the community workers engaged on the Bismarck-Ramu conservation project tried to rationalize their own method of dealing with indigenous knowledge, they said they were 'digging' and 'planting' (Ellis 1997: 45), not 'hunting' and 'gathering'. But they were still looking for knowledge of the 'natural forest', and what they seem to have found is knowledge of 'bush spirits', a form of pagan religious belief, rather than the sort of ethnoscientific knowledge which could usefully be sold to foreign scientists or ecotourists. They might well argue that the separation of these two forms of knowledge reflects the alienation of the landowners from their original social and intellectual landscape, or the transformation of a disenchanted forest into an economic resource. But arguments like this hold little water with the modern Christian villager who thinks of 'custom' as the work of Satan. Since most members of the national conservation community, unlike the anthropologists and other foreign scientists with whom they work, are active members of some Christian church, they cannot readily deal with the topic of indigenous knowledge without attempting to reconcile Christian and indigenous cosmologies. This kind of syncretism might appeal to Catholics or Anglicans, but is regarded with suspicion or hostility by most of the other churches.

The designers of the Bismarck-Ramu project tried to address this problem by engaging a Church Liaison Officer to liaise with local Christian leaders who might otherwise resist the traditionalism of the community development process. This man justified the significance of his own role by commending the ability of local pastors to enter an unfamiliar community and establish their own authority within a matter of days, by simply commanding the villagers to perform essential 'public work' (Ellis 1997: 40), and certainly without spending a lot of time 'hearing community themes' or exploring indigenous knowledge. No lessons seem to have been drawn from the contrast between the commanding presence of Christian preachers and the uncommanding, non-preaching talk favoured by the barefoot conservationists. But if we compare the Bismarck-Ramu project's style of community entry with the hard-nosed business preaching favoured by the Biodiversity Conservation Network, we may well be reminded of the different approaches which orthodox and evangelical missionaries took to the task of 'moulding Christian conviction', rather than 'conservation conviction', when they first arrived in Melanesia (Michael Young, personal communication). This does not mean that we are witnessing the spread of a new sect devoted to the worship of Biodiversity; it only means that conservationists and missionaries face a similar choice of moral incentives and rhetorical devices when they seek new converts to their cause. In both cases, the risk of appealing to indigenous knowledge is the risk of ceasing to be a 'true Christian'.

But this risk is greater for those conservationists who espouse the new Catholicism of the Bismarck-Ramu project than for those who espouse the BCN's version of the protestant ethic and the spirit of capitalism. That is because the 'Protestants' are offering material incentives, not just moral ones, integrating conservation and development by trading or exchanging the prospect of business development for the conservation of natural commodities, whereas the 'Catholics' reject developmentalism as a 'cargo cult', as yet another form of religion, and seek instead to cultivate the plant called self-reliance through the exchange of ideas, not the sale or purchase of commodities. The difference between them stems from their evaluation of the chances of successfully competing with 'developers', and most especially with logging companies, who promise more material benefits for less hard work.

The business experiments promoted by the BCN have not required a positive evaluation of traditional cosmologies. On the contrary, they look more like a set of minor rituals in the religion of developmentalism. But indigenous knowledge is still conceived as a sort of lost cause, in the sense that the current economic activities of the average villager, especially those which earn money, are seen to be minor versions of the menace posed by logging companies. The knowledge of animal behaviour that grew from traditional hunting practices has been distorted by the use of shotguns to satisfy new markets for particular species, while the knowledge of plants that grew from traditional cultivation practices has been distorted by the clearance of old forests for new cash crops (West 2000). While business is a good thing in itself, most existing forms of business are bad for biodiversity, and even the traditional arts of subsistence no longer carry the traditional guarantees of sustainability.

Yet those forms of indigenous knowledge that seem to have some bearing on the development of alternative forms of biodiversity business are also distorted through their attachment to economic activities and social relationships which are commonly less 'traditional' and less rewarding than the ones which they are meant to replace. It is one thing for the hunter to know his birds or the gatherer to know her trees, but such knowledge is not sufficient, and might not even be necessary, for people to make a decent living as small-scale saw-millers or guides on hire to adventure tourists. Nor is it obvious, either to landowners or economists, and even with substantial donor subsidies, how such occupations can generate the levels of cash income that would dissuade people from selling their birds or their trees, or persuade a typical community of Melanesian gardeners that an unexploited forest has more value than a garden full of coffee.

Whether we take the sacred or secular road to indigenous knowledge, conservation demands more time and effort, more education and collaboration, more conversation and negotiation, than local landowners are normally willing to supply, because conservationists tend to put their projects in places which already possess a natural abundance of biodiversity values, and often little else. In circumstances such as these, the conservationists are often selling victory against an enemy who is not really present in the field, and might never get there. In which case, we might ask why any package of incentives is required to fight a phantom?

> Conservationists are easily perceived as people who wish to maintain the status quo, which means poverty and a lack of services, by telling communities not to obtain 'development' by the only means that seem to be available. The populations of the Bismarck Fall and the Ramu Valley, for example, have, until now, lived under 'conservation circumstances'. They own a virtually untouched and enormous swathe of land in the Bismarck Fall, and to their great regret, experience a situation of 'conservation' rather than 'development' every day of their lives. It is unlikely that people will continue to accept this status quo if they are offered an alternative through mining, logging, or any other form of resource exploitation.
>
> (van Helden 1998a: 255)

The conservation of biodiversity values in a country like PNG is not necessarily, or even normally, the result of talks or deals by which conservationists persuade local landowners to change their attitudes or behaviour. Where the threats posed by developers exist only in the minds of landowners as a sign of their desperation for development, it does not follow that the current economic practices of those who labour under such illusions pose a smaller or a more insidious version of this threat. Where logging or mining companies really have made an assault on biodiversity values, with or without the active encouragement of local landowners, the latter may still combine their enthusiasm for development with demands for compensation which actually serve to undermine the object of their own desire, and conservation may then be the unintended or perverse result of their attempt to capture larger rents from the developers (Filer 1994). And whether the developers are present on their land or only present in their minds, local landowners may reduce their own exploitation of local forest resources, despite rapid population growth, because they want to modernize their own lifestyles, and thus dissociate themselves from those activities, like hunting, which are hallmarks of the forest-dwellers who they do not want to be (Kocher Schmid and Klappa 1999). The moral imperatives embedded in these options cannot be readily accommodated by the contrast between a valid defence of indigenous knowledge and a blanket condemnation of development-dependency.

If conservation biologists like to think of indigenous knowledge as knowledge about biodiversity, the province of the parataxonomist who shares their own interest in the classification of wildlife, we anthropologists are more inclined to think about it as a form of cultural diversity, and even to defend and celebrate it on these grounds. But if Melanesian biodiversity values have now become a sort of metaphorical commodity on sale to the donor community, what is the current value of Melanesian cultural diversity in any but the rapidly contracting market for traditional ethnography? And how is the conservation or reduction of this cultural diversity connected with those projects and policies that are funded and constructed for the purpose of sustaining its biological equivalent? The lingering diversity of traditional culture would seem to constitute an obstacle to the conservation of biodiversity because it adds an element of uncertainty to the reaction of different local communities to the community awareness or development strategies adopted by the conservationists. On the other hand, some communities appear to regard the conservation of their own cultural identity as an activity that makes more sense than the conservation of any particular species or habitat within their territorial boundaries (van Helden 1998a: 261). Since identities and boundaries are closely intertwined, their simultaneous defence may be the only point at which local landowners can accept a modern conservation ethic as something more than a luxury which they cannot and will not want to afford until they have some more 'development' (ibid: 263).

When we think of indigenous knowledge as the property of a local community, or even the cultural unity of a culture area, we also need to remember that natural biodiversity values are inversely related to population density, and that effective conservation of these values needs a space that is normally much greater than the territory occupied by a traditional political community in Melanesia. This means that conservationists tend to end up in places which are not only 'remote' and 'backward', but which also feature traditional cultural fault-lines, often overlaid by recent population movements, where talk of boundaries is highly problematic and political, and any kind of project which distributes benefits to local people can expect to start a mass of territorial disputes. The Bismarck Fall and the Lakekamu Basin are two such places (see Kirsch 1997; van Helden 1998a). A recent gathering of PNG 'ICAD Practitioners'

> noted that in the past there were no land disputes. Now that we have money there are disputes. The possibility that ICADs may exacerbate disputes should therefore be taken into account.
>
> (Saulei and Ellis 1998: 216)

Perhaps the traditional absence of land disputes belongs in the same romantic bag as the 'traditional conservation ethic', but their diagnosis of the present situation has some merit.

While this may seem to justify the Bismarck-Ramu project's method of withholding money and promoting 'self-reliance', its resource mapping practices could hardly fail to turn local knowledge into the knowledge of disputed territory. And if it is true that social relationships within and between Melanesian communities are normally grounded in the landscape (Filer and Sekhran 1998: 31), this may tell us more about the problem of striking 'conservation covenants' with local landowners than about their understanding of a conservation ethic or their appreciation of indigenous knowledge.

> Where several local communities are given the time and space in which to make their own choice between activities which may have positive or negative impacts on biodiversity values, there is no guarantee that people who make the 'right' choice will occupy a series of contiguous territories which combine to make a single conservation area of several thousand square kilometres. If anything, it is rather more likely that neighbouring communities, or even sections of a single community, will take different options as a new means to represent and pursue their existing social and political divisions.
>
> (ibid: 294–296)

The conservationist's dilemma here is that the weaker party generally takes the side of conservation, but the weaker party typically lacks the numbers and capacity to protect and manage a conservation area without external assistance. One of the most striking features of the current Melanesian form of landed property is that the power of local landowners to exploit 'projects' for their own ends is not matched by their power to deal with troublesome neighbours, nor is it matched by the power of the state to keep local boundaries in order.

At the end of the day, as we still beat around the bush, we need to ask whether conservationists who often seek to avoid a confrontation with developers, which they cannot hope to win, can win the hearts and minds of landowners whose own achievement of conservation is not exactly what they want. The Occidental donors who fund conservation projects are not in the business of funding the roads, schools or other public goods that would satisfy the most immediate and explicit desires of the rural population. And that is partly why the mediators in the conversation between Western conservationists and Melanesian landowners are backed into a corner where indigenous knowledge gains its peculiar significance. But the celebration of indigenous knowledge by community development strategies that are premised on the existence of a traditional conservation ethic, or the partial exploitation of such knowledge by business development strategies which seek to divert rural villagers from their current economic practices, cannot readily address the basic facts of rural poverty and social inequality.

For this reason, we also need to ask whether conservationists could achieve more of what they want by seeking to modify the behaviour of local communities than they would achieve by either doing nothing at all, leaving landowners alone to suffer in silence, or by using their own scarce resources to change the actions of other characters in the policy process (Dove 1996; Brandon 1997). I have previously argued that the Occidental donors could carry on their fight with the Oriental loggers by the simple expedient of persuading the national government to cease the dispensation of timber permits in areas which are known to be rich in biodiversity, without necessarily having to fund conservation projects in those areas (Filer and Sekhran 1998: 257–258). The World Bank has actually used its own financial leverage to wrap increasing volumes of red tape around a forest industry which it now regards as a threat to the values of good governance (Filer, Dubash and Kalit 2000), and in this respect, it has helped to create the physical space in which conservation projects are freed from the necessity of competing directly with the logging companies. But if the conservationists have bought more time in which to carry on their dialogue with local landowners, they may still find, at the end of the day, that money talks louder than birds.

References

Allen, B.J. 1986. Environmental ethics and village agriculture. *Yagl-Ambu* 13 (3): 1–14.

Ballard, C. 1997. 'It's the land, stupid!' The moral economy of resource ownership in Papua New Guinea. In *The Governance of Common Property in the Pacific Region.* (ed) P. Larmour. Pacific Policy Paper 19. 47–65. Canberra: Australian National University, National Centre for Development Studies.

BCN (Biodiversity Conservation Network). 1997. *Getting Down to Business: Annual report 1997.* Washington, DC: BCN.

—— 1999. *Evaluating Linkages Between Business, the Environment and Local Communities: Final Stories from the Field.* Washington, DC: BCN.

Brandon, K. 1997. Policy and practical considerations in land-use strategies for biodiversity conservation. In *Last Stand: Protected areas and the defense of tropical biodiversity.* (eds) R. Kramer, C. van Schaik and J. Johnson. New York: Oxford University Press. 90–113.

Brosius, J.P. 1999. Analyses and interventions: Anthropological engagements with environmentalism. *Current Anthropology* 40: 277–309.

Bulmer, R. 1982. Traditional conservation practices in Papua New Guinea. In *Traditional Conservation in Papua New Guinea: Implications for today.* (eds) L. Morauta, J. Pernetta and W. Heaney. Monograph 16. 39–77. Boroko: Institute of Applied Social and Economic Research.

CRI (Christensen Research Institute). 1996a. Report of the First Two Community Development Patrols in the Bismarck-Ramu ICAD Area of Interest. Unpublished report to the PNG Conservation Resource Centre.

——1996b. Report of the Second Two Community Development Patrols in the Bismarck-Ramu ICAD Area of Interest. Unpublished report to the PNG Conservation Resource Centre.

—— 1996c. Report of the Third Round of Community Development Patrols in the Bismarck-Ramu ICAD Area of Interest. Unpublished report to the PNG Conservation Resource Centre.

—— 1997a. Report of the Fourth Round of Community Development Patrols in the Bismarck-Ramu ICAD Area of Interest. Unpublished report to the PNG Conservation Resource Centre.

—— 1997b. Report of the Seventh Round of Community Development Patrols in the Bismarck-Ramu ICAD Area of Interest. Unpublished report to the PNG Conservation Resource Centre.

Croll, E. and D. Parkin. 1992. Cultural Understandings of the Environment. In *Bush Base, Forest Farm: Culture, environment and development.* (eds) E. Croll and D. Parkin. 11–36. London: Routledge.

Diamond, J. 1997. *Guns, Germs and Steel: A short history of everybody for the last 13,000 years.* London: Jonathan Cape.

Dove, M.R. 1996. Center, periphery, and biodiversity: A paradox of governance and a developmental challenge. In *Valuing Local Knowledge: Indigenous people and intellectual property rights.* (eds) S.B. Brush and D.F. Stabinsky. Washington, DC: Island Press. 41–67.

Dwyer, P.D. 1982. Wildlife conservation and tradition in the highlands of Papua New Guinea. In *Traditional Conservation in Papua New Guinea: Implications for today.* (eds) L. Morauta, J. Pernetta and W. Heaney. Monograph 16. 173–189. Boroko: Institute of Applied Social and Economic Research.

—— 1997. Modern conservation and indigenous peoples: In search of wisdom. *Pacific Conservation Biology* 1: 91–97.

Ellen, R.F. 1993. Rhetoric, practice and incentive in the face of the changing times: A case study of Nuaulu attitudes to conservation and deforestation. In *Environmentalism: The view from anthropology* (ed.) K. Milton. ASA Monograph 32. 126–143. London: Routledge.

Ellis, J.-A. 1997 [actually published 1999]. *Race for the Rainforest 2: Applying lessons learnt to the Bismarck-Ramu integrated conservation and development process in Papua New Guinea.* Port Moresby: PNG Biodiversity Conservation and Resource Management Programme.

Filer, C. 1991a. The eyes of the world are on Lak. *Times of PNG.* 3 and 10 January.

—— 1991b. Two shots in the dark: The first year of the task force on environmental planning in priority forest areas. *Research in Melanesia* 15(1): 1–48.

—— 1994. The nature of the human threat to Papua New Guinea's biodiversity endowment. In *Papua New Guinea Country Study on Biological Diversity.* (eds.) N. Sekhran and S. Miller. Waigani: PNG Department of Environment and Conservation. Nairobi: Africa Centre for Resources and Environment. 187–199.

—— 1997. Compensation, rent and power in Papua New Guinea. In *Compensation for Resource Development in Papua New Guinea.* (ed.) S. Toft. Monograph 6. Boroko: Law Reform Commission. Canberra: Australian National University, National Centre for Development Studies (Pacific Policy Paper 24). 156–189.

Filer, C. with N. Sekhran. 1998. *Loggers, Donors and Resource Owners.* London: International Institute for Environment and Development in association with the National Research Institute (Policy That Works for Forests and People, Papua New Guinea Country Study).

Filer, C. with N.K. Dubash and K. Kalit. 2000. *The Thin Green Line: World Bank leverage and forest policy reform in Papua New Guinea.* Monograph 37. Boroko: National Research Institute. Canberra: Australian National University, Research School of Pacific and Asian Studies, Resource Management in Asia-Pacific Project.

Filer, C., M. Hedemark, A. Kituai, S. Majnep and C. Unkau. 1995. *Ramu Conservation Area: April 1995 site visit.* Unpublished report to the PNG Conservation Resource Centre.

Filer, C. and W. Iamo. 1989. *Base-Line Planning Study for the Lakekamu Gold Project.* Unpublished report to the PNG Department of Minerals and Energy.

Grant, N. 1996. *Community Entry for ICAD Projects* – The participatory way. Port Moresby: PNG Biodiversity Conservation and Resource Management Programme.

Grillo, R.D. and R.L. Stirrat. (eds) 1997. *Discourses of Development: Anthropological perspectives.* Oxford and New York: Berg.

Kirsch, S. 1997. Regional dynamics and conservation in Papua New Guinea: The Lakekamu River Basin Project. *Contemporary Pacific* 9 (1): 97–121.

Kocher Schmid, C. and S. Klappa. 1999. Profile of a leader, or the world according to Yulu Nuo. In *Expecting the Day of Wrath: Versions of the millennium in Papua New Guinea.* (ed.) C. Kocher Schmid. Monograph 36. Boroko: National Research Institute. 89–109.

Lalley, B. 1998. Is conservation the answer? In *The Motupore Conference: ICAD practitioners views from the field.* (eds) S.M. Saulei and J.-A. Ellis. Waigani: PNG Biodiversity Conservation and Resource Management Programme. 16–19.

Margoluis, R. and N. Salafsky. 1998. *Measures of Success: Designing, managing and monitoring conservation and development projects.* Washington, DC: Island Press.

McCallum, R. and N. Sekhran. 1997. *Race for the rainforest: Evaluating lessons from an integrated conservation and development experiment in New Ireland, Papua New Guinea.* Port Moresby: PNG Biodiversity Conservation and Resource Management Programme.

Nelson, H. 1976. *Black, White and Gold: Goldmining in Papua New Guinea, 1878–1930.* Canberra: Australian National University Press.

Orsak, L. 1997. *Twenty Arguments for NOT Using Natural Resources Sustainably and the Weaknesses with these Arguments: Villager perspectives in Papua New Guinea.* Madang: Christensen Research Institute.

Saulei, S.M. and J.-A. Ellis (eds). 1998. *The Motupore Conference: ICAD practitioners views from the field.* Waigani: PNG Biodiversity Conservation and Resource Management Programme.

Van Helden, F. 1998a. Between cash and conviction: The social context of the Bismarck-Ramu Integrated Conservation and Development Project. Monograph 33. Boroko: National Research Institute.

—— 1998b. *The Bismarck-Ramu Social Feasibility Study: Overview and recommendations.* Unpublished report to the PNG Biodiversity Conservation and Resource Management Programme.

—— n.d. The Community entry approach of the Bismarck-Ramu Integrated Conservation and Development Project. To be published in *Custom, Conservation and Development in Papua New Guinea.* (ed.) C. Filer. *Forthcoming.*

West, C.P. 2000. *The Practices, Ideologies, and Consequences of Conservation and Development in Papua New Guinea.* New Brunswick, NJ: Rutgers University (PhD thesis).

World Bank. 1991. Operational directive 4.20: *Indigenous Peoples.* Washington, DC: World Bank.

Young, M. 1991. Logging or conservation on Woodlark (Muyuw) Island. *Research in Melanesia* 15: 49–65.

Zerner, C. 1996. Telling stories about biodiversity. In *Valuing Local Knowledge: Indigenous people and intellectual property rights.* (eds) S.B. Brush and D.F. Stabinsky. Washington, DC: Island Press. 68–101.

Chapter 6

Close encounters of the Third World kind

Indigenous knowledge and relations to land

Veronica Strang

This chapter considers the relationship between systems of knowledge and attachment to land. Examining ethnographic data from an Aboriginal community in North Queensland and Euro-Australian pastoralists on the surrounding cattle stations, it argues that the use of land as the primary medium for the location of cultural knowledge engenders 'place-based' identity and affective environmental relations which are not experienced to the same degree by more transient cultural groups. Implicit in this argument is an assumption that indigenous knowledges and identity have specific characteristics and are located in 'place' in ways that are meaningfully different to the more fluid knowledge and identity constructions of other societies.

Environmentalists have often represented indigenous groups as ideal models, not only of 'harmony with nature' but also of social and emotional coherence, and continuity. Anthropology has tended to reject these romantic images. However, the data presented here suggest that there are real differences in indigenous knowledge systems that provide a powerful rationale for their use as ideal models of sustainable resource management and environmental values which integrate human and ecological needs.

To examine the characteristics that may be said to define indigenous knowledge, first we must consider how these enable cultural groups to consider how these enable cultural groups to construct localised identities and communities, often in opposition or resistance to a globalising social and economic environment. Focusing on an Aboriginal community in northern Australia and the Euro-Australian pastoralists in the same area, it argues that the use of land as the primary medium for the location of cultural knowledge engenders a form of integrated knowledge,[1] place-based identity and affective relations with a specific landscape which are not experienced to the same degree by more transient cultural groups. Implicit in this argument is the assumption that indigenous knowledges and identities are inherently local – intimately bound up with place in a way that is significantly different to the knowledge and identity constructions of other societies. In effect, it is not tenable to consider an indigenous system of knowledge unmediated by a specific landscape.

Defining this significant difference is an intensely political issue both in terms of land rights and with regard to development. As Ellen points out (citing Street 1975), environmentalists, romantics and alternative groups, have tended to present indigenous peoples – and particularly hunter-gatherers – as model societies which have an almost magical harmony with nature (Ellen 1986). Aboriginal Australians and Native Americans have long provided inspirational images of people at home in the wilderness, spiritually attuned, wise about ecology, and deeply attached to their traditional lands. Somehow, this comparison implies, these groups have proper social and environmental values, that Westerners, caught up with material production and consumption, have forgotten. Though the stereotype has been politically advantageous for some minority groups, most particularly in negotiations about development, such reification of indigenous knowledge requires examination. In recent years, it has become fashionable to debunk this romantic vision, and to suggest that, given half a chance (i.e. with sufficient technology) indigenous groups would be – or would have been – as venal and as destructive to the environment as any industrialized society. In Australia, amid the pressures for rapid development and the tensions generated by increasing conflicts over land, such debunking has a sharp political edge, with groups opposing Aboriginal land rights and/or the green movement all too keen to claim that indigenous management has wrecked the environment and decimated species, and that Aboriginal attachment to land is nothing more than a cynical scam to gain control of large areas of it. At the same time, they say, white Australian landowners have an attachment to land that is the same – or equal to – that of Aboriginal people, and therefore just as much right to it.

In some respects the sceptics have a point: Aboriginal relations with land have been considerably romanticized and are, in reality, as pragmatic as they are spiritual. Aboriginal land management over 60,000 years has probably had major long-term effects on the Australian landscape. No doubt it is also fair to say that European Australians have an equal capacity for attachment to land, and there is plenty of evidence (e.g. Read 1996; Strang 1997) to suggest that many have indeed constructed strong relationships with particular places through processes of bonding engendered by the investment of labour, continuous interaction with the land, spiritual and aesthetic appreciation, and the creation of special places for burial, recreation and so on.

Anthropologists have contributed to the demystification of indigenous knowledges by attempting to articulate the complex processes through which all human beings engage with their environments. Their efforts have generated sophisticated concepts of landscape that consider dynamic cultural constructions of place (e.g. Bender 1993; Tilley 1994; Hirsch and O'Hanlon 1995), and a range of more abstract models of human-environmental interaction (e.g. Morphy 1998; Ingold 1995, 2000; Bourdieu 1977). This has been a useful discourse within anthropology, but it has tended to obscure rather than resolve the question as to whether local knowledges and identities are meaningfully different from those constructed within a global milieu.

It is logical enough to suggest that local and global may be considered as the polarities of a continuum, and that people situate themselves along this, constructing spheres of knowledge with varying degrees of locality and abstraction based on their particular socio-economic networks, opportunities and access to local/global media. Though this is reasonable, it homogenizes the processes of knowledge and identity creation, and therefore does not elucidate what may be critical differences in indigenous knowledge systems. Nor does it explain why these are reified and romanticized as being fundamentally different from so-called Western types of knowledge.

This chapter argues that there are real differences in indigenous knowledge systems, and that these provide a powerful rationale for their use as models of ideal environmental relationships. In essence, it suggests that the holistic location of all aspects of cultural knowledge in an immediate physical landscape permits the integration of social and environmental relationships and an intimacy of connection between people and place that is psychologically satisfying. This close correlation of knowledges, and the use of the physical landscape as a primary medium, ensures that each area of knowledge is continually reaffirmed and supported, and thus the creation of place-based knowledge and identity is more than the sum of its parts or, as Gurung puts it, more than a sum of its 'utility functions' (1997:19). Further, it suggests that the compartmentalized, multi-layered and fragmented types of knowledges constructed in mobile, large-scale societies are intellectually difficult to integrate and thus socially and psychologically unsustaining. Berger *et al.* presciently described this crisis of modernity as intellectual homelessness (Berger, Berger and Kellner 1973). Implicit in their argument is the idea that such forms of knowledge are lacking in place. If this is so, it is unsurprising that many Westerners, observing the rootedness of local or indigenous knowledge and identity, yearn for the security of its assumed social and environmental certainties.

There are echoes here of earlier anthropological writing about 'open' and 'closed' societies, which framed indigenous cultural groups as being largely closed. These terms seem slightly pejorative now, in a society where openness is generally regarded in a positive light. However, what does emerge from these ideas is a sense of differing shapes of knowledge, some of which are genuinely more open to radical change than others. This chapter attempts to elucidate some of the arguments outlined above by comparing two very different shapes of knowledge. Drawing on ethnographic research with a range of land using groups in the Mitchell River area of Cape York, North Queensland, it considers how Aboriginal people in Kowanyama construct a system of what is clearly highly localized knowledge, and how they use their immediate landscape as the primary medium for all of its aspects. It then compares this with the way in which cattle farmers in the surrounding area construct their own systems of knowledge.

Home is where the heart is

The community in Kowanyama is composed of about 1,200 people: three different language groups who, though their lands were more widespread prior to the European invasion, now live in an area of about 1,000 square miles, on the coast of the Gulf of Carpentaria. It is surrounded by vast cattle stations, the majority of which were established during the gold rushes of the late 1800s. Kowanyama was originally a church mission: formed in 1905, the settlement and the surrounding land were handed over to the state in the 1960s, and it is now Aboriginal land, having been granted to the community formally in 1987. Before the recent tourism boom in the far north, it was defined as a remote area, being many hours by dirt road from the more populated East coast.

For much of the twentieth century its inhabitants have therefore remained relatively isolated from the mainstream, although there was steady contact with missionaries, cattle farmers and various government officials. Christianity was, rather patchily, laid over Aboriginal beliefs, and members of the community were employed as domestic or stock workers on the neighbouring cattle stations, or in Kowanyama as shopkeepers, cleaners, clerical workers and local councillors. Following its mission days, like most such communities, Kowanyama became increasingly dependent on economic support from the state and subject to its educational, medical and legal systems.

So there has been no shortage of outside influences on the Aboriginal community, and these have increased dramatically in the last 20 years, as Cape York, once a forgotten backwater, has become a developer's paradise and a desirable wilderness experience for tourists. Despite these pressures, however, and the inevitable changes that they represent, the community in Kowanyama, like many other indigenous groups, has demonstrated great resilience in maintaining Aboriginal knowledge and identity. Though semi-dependent on the state, people supplement this economic support by hunting and gathering. They maintain many of their own religious practices and look to their elders for gerontocratic leadership. Kinship provides the basis of social, economic and spatial organization. Thus the traditional beliefs and values documented by anthropologists over 50 years ago (see Sharp 1937, 1939, 1952) remain central to the community's activities and discourses, and people still express – loudly and clearly – their deep attachment to land. Today, with greater political self-determination, groups in the community are attempting to reclaim their land and regain economic self-sufficiency, while explicitly trying to do so in a way which reflects their own rather than imposed European values.

One of the key questions with regard to indigenous knowledge is how, despite the massive dislocations caused by colonial dominance and increasingly intense modern pressures for change, groups such as the Aboriginal communities in Australia have maintained such robust cultural continuities. In Kowanyama, the cultural practices described in Sharp's ethnography (ibid) remain central to contemporary life. The foundation of this resilience is the land: for every Aboriginal group, even those who have been relocated to urban areas, land remains central to constructions of knowledge and identity. Furthermore, the ethnographic data suggest that this centrality is a constant factor in every area of traditional knowledge.

Much has been written about the spiritual aspects of Aboriginal culture (e.g. Durkheim 1954 [1912]; Charlesworth *et al.* 1984; Morphy 1988; Taylor 1974; Layton 1989), but it is useful to reprise a few key points here. In an Aboriginal cosmology, the land is not merely a material surface: it was formed in the Dreamtime by ancestral beings, deities in the forms of animals and birds, who emerged or became visible, created all the features of the landscape, all its inhabitants, and so forth. Each ancestral myth – or song – describes these journeys in detail: what the ancestral being did, where she or he went, all the events of their travels. Having completed their journeys, the ancestors sat down into the land – i.e. shifted back to an invisible plane of existence – where they remained, creating a living, sentient landscape that watches and responds to human action and provides a source of spiritual power. Held in the land, the various ancestral beings became totems[2] for the different human clans who, in Aboriginal terms, inherited their country from these ancestors, and therefore own it in perpetuity.

According to Aboriginal beliefs, human spiritual being emerges from these ancestral forces, and every individual has – usually within their own clan land – a particular place from which their spirit has emerged. More often than not this is a water source of some kind, and the spirit children held in the place jump up, as Aboriginal people say, to give the spark of life to the foetus in a woman's womb. In this way, every human being emerges from a particular place, a spiritual conception site which they call their home or in *Kunjen* (one of the three major Aboriginal languages spoken in Kowanyama), their *errk elampungnk.*[3] Like the ancestral beings, people come out of the land, grow and mature. Acquiring spiritual knowledge as they age, they become more valued and venerated, coming closer to their ancestors. When they die, their spirit is ritually returned to its home, to become one with their clan totem, to be regenerated in a subsequent generation.[4] So every human life reflects the ancestral cycle of emergence onto a visible plane of existence, and eventual reintegration, into the invisible, spiritual dimension held in the land. Thus the land provides the very basis of human 'being', imbuing it with spiritual meaning and emotive force.

As well as providing a cosmological explanation of creation and human spiritual genesis, this densely metaphorical account of the world also constitutes social knowledge. Because in Aboriginal terms, an individual's ancestral home is the basis of their social identity, if their conception site is on Emu clan land, then they are, inextricably, part of Emu clan, part of the Emu Story and – most importantly – part of this particular landscape. Each of these clans is part of a wider exchange network. In the pre-colonial period marriages were arranged according to an ideal model defined by ancestral Law,[5] and people were supposed to marry classificatory cross cousins.[6] The social and physical landscape was composed of a mosaic of clan lands, with people (and artefacts) circulating

Figure 6.1 Dance recounting an ancestral myth in Kowanyama.

steadily around this network, within a larger language group. At the margins, exchanges would be made with other language groups,[7] and thus Australia was inhabited, with varying degrees of density, by a vast network of Aboriginal clans, linked through exchanges of people and material culture. The major point about this system is that each of these relationships is defined more by geography than genealogy. As indicated by other ethnographic accounts (e.g. Morphy 1988; Hamilton 1982), social knowledge is shaped and mediated by the land, and land is therefore, simultaneously, a cosmological map of ancestral forces, and a map of people and their social relationships. For Aboriginal groups kin and country are indivisible: it is literally impossible to talk about one without referring to the other and in this way land represents the emotional, affective aspects of people's lives, embodying their kin relationships and tying their personal lives into other forms of knowledge. This conflation leads to an intensity of affective attachment that is difficult for non-Aboriginal people to appreciate. The closest analogy for Europeans might be a cemetery in which all of their ancestors, including their nearest relatives, are buried, but in the case of Aboriginal land, the country holds not only relatives who have died, but also all of those who are alive, and those yet to come. The loss of country is thus also the loss of familial connections, and the elders relate how Aboriginal people sent from Kowanyama to a prison on Palm Island cried for their country and tried to evoke it by singing its ancestral songs.

These powerfully affective forms of knowledge, located in the land, create a deep cathection[8] with place. This is supported and enabled by highly practical forms of knowledge: tied into this socio-spatial body of knowledge is an entire economic system for which land is similarly foundational. As noted previously, land is owned collectively by the clan, and membership of a clan entitles people to inalienable ownership of particular tracts of land and their resources.[9] Under Aboriginal law, this is a secure and permanent relationship. Before the Europeans arrived, people lived in small family groups for much of the year, hunting and gathering in regular patterns around their clan land, making use of all of the local environment's different resources – gathering seeds, nuts, berries and shellfish, using bush medicines, making a few (largely portable) artefacts and fishing in the waterholes and rivers. In the wet season they made shelters and stayed up above the floods on the sand ridges, and in the dry they travelled more, congregating for annual social events, communal hunting and fishing, exchanges, religious rituals, marriages and suchlike. Occasionally individuals traded goods further afield, but this required special permission from other groups, and in general clans maintained a steady annual cycle of movement around their own land. The archaeological evidence suggests that Aboriginal Australians maintained this sustainable economy, with great continuity, for over 60,000 years. In recent years, through involvement in stock work and tourism, through welfare support, and the introduced services within a state-run reserve, the inhabitants of Kowanyama have made linkages with wider economic systems. However, this remains secondary in many respects, and where people

Figure 6.2 Hunting and gathering in contemporary Aboriginal communities.

do have access to other resources (such as vehicles or wages) these are managed in accord with their particular kin obligations and local economic practices.

As many ethnographers of hunter-gatherers have established (e.g. Ingold, Riches and Woodburn 1988; Altman 1984; Jones 1990; Williams and Hunn 1986; Sutton 1978) a successful hunting and gathering economy is necessarily reliant on certain kinds of knowledge and skill. More than anything else, it requires a detailed understanding of all the potential resources within the local environment: knowing not only how to recognize the species that are either useful of poisonous, but also having intimate knowledge of their seasonal changes or habits and, perhaps most crucially, knowing exactly where and when they can be found within a vast and often harsh landscape. It requires, in other words, an immensely detailed lexicon of ecological knowledge about a specific landscape. Such knowledge is acquired through continual, practical engagement. Hunting and gathering is the most intimate and holistic kind of interaction with a landscape that human beings can have, requiring physical and intellectual engagement with every part of the local ecology: constant observation, close sensory contact and careful deduction. It demands a continual and intense focus of attention on the land which is only rarely required in most economic modes.

Hunting and gathering also requires people to stay put. Although it has been popularly suggested – with critical political and economic consequences – that Aboriginal Australians merely wandered about all over the place, in fact their movements were largely confined to their own (sometimes very large) tracts of land, not only because this is where they had rights to resources, but precisely because such knowledge is essentially local.[10] Though people with good ecological knowledge and well-honed observational and recall skills can and do learn new places very quickly when forced to move, hunting and gathering is most efficiently carried out when people know exactly where resources are, and can count on them being there. So Aboriginal ecological knowledge, like spiritual and social knowledge, is firmly located within a particular landscape.

All of these aspects of cultural knowledge are neatly woven into the ancestral myths, which in Kowanyama (as elsewhere in Aboriginal Australia) are referred to as the Law. The ancestral stories are holistic, encompassing every aspect of Aboriginal life: they delineate clan country and kin relations through an account of the totemic beings, and describe rituals, responsibilities, social roles and rules of behaviour. Through ancestral parables they deal with issues of law and morality. They contain oral maps of the local topography, nicely described in the travels of the totemic beings; and they are full of details about different species and resources, their recognizable characteristics, how to catch or gather them, how to cook food or construct artefacts. In essence the ancestral stories are a blueprint for a whole way of life. This is transmitted inter-generationally through a rich oral tradition, through paintings, sculptures, other storytelling artefacts, and in dances, songs and rituals (e.g. Morphy 1991; Munn 1973, 1984; Strang 1997, 1999).

There are two major points to make about this body of indigenous knowledge: first, that it is all – every part of it – written into the land and, second, that this location in place serves to integrate each of its strands. Because the land itself is the primary medium and repository of knowledge, each area of knowledge is so closely woven into the whole that one cannot refer to any part of it without in some way referring to the rest. Numerous ethnographic accounts (ibid) have shown that Aboriginal knowledge is firmly 'held in place' and closely conflated in conceptual terms. Thus, as Wilson notes (1988: 50, cited in Ingold 1995) 'the landscape is turned into a mythical topographic map, a grid of ancestor tracks and sacred sites'. So people's experience of being 'in place' is not merely a matter of going there to get resources, but is a way of engaging with knowledge about kin relations, spiritual and emotional life, morality and so forth. Thus the land becomes a repository for the history of groups and individuals, which in Aboriginal terms goes back indefinitely, or as one woman put it, 'to the beginning to us' (Alma Wason).

Ethnographic investigation of indigenous knowledges in Kowanyama points to other key issues. For example, Aboriginal discourses are dominated by spatial rather than temporal metaphors, and are based on cyclical rather than linear concepts of time. As various writers have noted, (e.g. Gosden 1994; Gould 1987) this is an attribute shared by many indigenous cultures. According to Aboriginal law, human lives echo ancestral journeys in their emergence from the Dreaming and their eventual reintegration with the ancestral forces, and, ideologically, human beings are required to relive the lives of the ancestral beings. It is difficult to imagine a more conservative cosmological model, or one more designed to be resistant to change. Crucially though, this model of time also means that linear history is de-emphasized, while spatial meaning comes to the fore. As noted previously, (and by Munn 1986; Morton, 1987) the Dreamtime is merely another dimension that exists, invisibly, alongside – or perhaps one should say underneath – the present. So it is less a creative era, and more of a place where things happen. The dominance of spatial metaphors underlines the importance of place in the construction of indigenous knowledge and points to another important characteristic: that such knowledge is constructed on a scale that is very immediate. There is no separation of linear millennia from the gods: they are located right on the doorstep, powerfully manifested through ancestral creativity in a sentient landscape.

Another major characteristic, which also appears to be shared by other indigenous groups, is that discourses about the environment are primarily qualitative (see Rudder 1983; Myers 1986; Munn 1986). It is a truism that Aboriginal languages have no words for numbers above three, and informants in Kowanyama confirm that although it is possible in their own languages to make sets of three, more commonly more than three is just a lot. This qualitative emphasis is nicely illustrated by an ethnographic example: on the cattle station at Rutland Plains, just outside Kowanyama, it was regularly necessary during the mustering season to bring in all of the work horses (about 60 animals) from the paddock. Generally the head stockman would count them in, to ensure that all had been collected, but it became obvious, whenever any were missing, that this was not the methodology used by the Aboriginal stockmen. They would recognize each horse individually, and could invariably name and describe any that were absent. This highlights a further, related point about indigenous knowledge – that it is typically specific rather than generic. As anyone who has worked with Aboriginal Australians can attest, even the most general questions to informants will almost always produce a specific example in response. Mapping the ancestral myths underlines this specificity: all places are unique; all of the information located in them is particular. This too is an issue of scale: circumscribed by cognitive limitations, knowledges that are specific rather than generic, qualitative rather than quantitative, are necessarily limited in scale. It could be argued, conversely, that the management of larger systems of knowledge is reliant upon reductive cognitive mechanisms such as quantification and the creation of generic categories. More critically, with regard to indigenous knowledge, it is feasible to posit that specific, qualitative forms of knowledge depend upon the continuous use of a physical environment of mnemonics and the shaping of knowledge in such a way that each strand reaffirms another: for example as in the conflation of topographic, genealogical, economic and social knowledge. This suggests that the characteristics of indigenous knowledge and its location in place are interdependent.

What emerges from this sketch of an indigenous system of knowledge is a sense that it is, if not entirely closed, intensely local and self-reaffirming, and clearly well designed to resist external influences. It is therefore unsurprising that it has proved to be immensely resilient, although there are some important questions about how its integrity is affected by efforts to absorb knowledges and identity formulations that are not related to place. So far, however, where Aboriginal people have been forced – as they often have – to adopt other religious practices or economic modes, these are merely added as an overlay – a *bricolage* as Levi-Strauss called it (1968) – to an existing way of interacting with the environment. Thus, in Kowanyama, God becomes *M'atat*, a larger and more abstract overseer to the local ancestral beings, and cattle ranching extends the skills already honed by hunting other animals. Western technology is incorporated into existing economic obligations to specific relatives, and new roles, such as the creation of Aboriginal Rangers are framed according to traditional precepts (see Strang 1998).[11]

It appears, therefore, that anchoring cultural knowledge in the land enables indigenous groups to cathect thoroughly with their physical environment and thus to maintain some certainties and continuities that, from the perspective of urban Australians stressed by constant change and uncertainty, are somewhat enviable. Concomitant with this location of knowledge in a physical landscape is an extraordinarily close and intimate connection with place, leading to a sense of belonging and a depth of affective attachment which, for homeless Westerners, appears equally desirable. In an Aboriginal system of knowledge, all roads lead to home.

Home is where I hang my hat

> Home is where me hat is. That's where my home is . . . we're still . . . wandering around, sort of thing, but together, together. We haven't got anything . . . I mean like these days jobs aren't permanent are they? Things happen and you just don't know what's likely to go on (Diane Denial).

Quite different characteristics define the system of knowledge and the concepts of place held by the Euro-Australian pastoralists who live on the vast cattle stations surrounding Kowanyama. The pastoral sub-culture is central to Australian constructions of identity, an identity that is – in its own terms – validated by a history of pioneering, settlement and the investment of hard *yakka* in the land. As the right-wing One Nation Party states, this is 'vernacular Australia'. Since late in the last century (Cape York being settled much later than the southeastern seaboard), pastoralists have lived and worked on cattle properties ranging from 1,000 to over 3,000 square miles, often in areas that are more than 12 hours drive by dirt road from the nearest town or village.

The cattle industry has therefore been essential to the process of colonization, throwing an appropriative net across the landscape, dispossessing or dominating the indigenous landowners, and, by placing small clusters of Euro-Australians over the land, effectively establishing colonial authority. Like settlers everywhere, the pastoralists focused their energies on renaming the landscape, humanizing it in their own terms, bringing it under technological control, and making it economically productive.

Figure 6.3 Stockwork in North Queensland.

Their economic mode is highly specialized, but quite straightforward: having criss-crossed the landscape with fence lines that divide their properties into large paddocks, they disperse cattle in accord with the availability of water and grass, muster them annually using a combination of helicopters and horse riders, draft off a percentage for the beef market, and redistribute the remainder around the property. Little or no use is made of other local resources, though there is some recreational hunting and fishing, and occasional supplementation of income from cattle through the capture of feral pigs. This latter endeavour highlights the fact that, in Cape York, many stations are barely viable financially: the distance to markets is too great and the land is not well suited to raising cattle. Pastoralism has therefore become a marginal industry, supported more for its historic imagery and for its continued holding of the land than for any major contribution to the state economy.

In social terms, the pastoralist sub-culture was until recently drawn mainly from the rural population, offering work to the least educated echelons of Australian society. Many stockworkers had few literacy skills and limited employment alternatives. This is changing now: cattle work has become more of a rite of passage for middle class youths, and economic stringency has made it difficult for stations to hire experienced ringers. Despite these changes though, the pastoralists remain one of the most intensely conservative sub-cultural groups in Australia, demonstrating, through persistent right-wing political activism, a deep resistance to change.

Their interaction with the land is more intimate than that of most Australians, and thus most akin to that of the indigenous people. Despite this – or more likely because of it – the pastoralists are, in general, violently opposed to Aboriginal interests. In essence, they have been on the land longer than other Euro-Australians: stockwork brings them into continual engagement with the physical landscape – their days are spent hunting down and capturing animals – and they need some, albeit specialized, local knowledge to do this successfully. Superficially, it would be reasonable to expect that their knowledge and identity would share some of the characteristics of local indigenous knowledges. There is some similarity: for example, the pastoralists have a tighter and more stable social network than that of many Australians and it is, to some degree, located in place. Their colonial humanization of the landscape has left a legacy of names and histories of interaction in the land, and, in modern terms, people are still socially identified with the properties on which they live and work. Most managers and head stockmen have a considerable local knowledge about the topography of the cattle station, its water resources and the availability of feed for the cattle. Their reading and representations of the land are dominated by these features and the fence lines, roads, holding paddocks and yards that organize the property economically. They also have some – albeit sketchy – knowledge about the local flora and fauna. The more junior members of the stock team, due to the nature of their work, will have a particularly closely embodied knowledge of the physicality of the immediate environment – the prickly grasses and thick scrub, the garrotting rubber vine, the heat and the dust.

So there is some evidence that Gellner's 'potato principle'[12] is at work here – in settling on the land, the pastoralist sub-culture appears to have constructed local forms of knowledge and identity that have some characteristics in common with those of the indigenous community. However, there are some crucial differences which reveal that this is little more than a passing resemblance. A review of the ethnographic data shows that much of the pastoralists' knowledge is only superficially, partially or transiently located in place. The first point to be made is that their vaunted historical continuity is something of a romantic fiction. Though some cattle ranching families remained on their properties for several generations, and in comparison to urban Australians the pastoralists are a relatively settled group, this is actually a very mobile population. The most long-term residents on the cattle stations are owners and their families, but most station owners leave their properties to be managed by outsiders who remain – at the most – for a decade or two. The managers arrive at the properties as mature adults and invariably retire elsewhere. The stock team and domestic staff (governesses and cooks) are even more mobile, rarely remaining on the same property for more than a year or two. Many drift in and out of stock work, alternating this with other forms of employment such as mining or labouring. For many youths (who compose the majority of most modern stock teams) work on the stations is merely a brief rite of passage before they go on to other things. So while people may identify with the stations on which they work, this is a temporary persona, and even when they construct identities in the longer term as 'cattle folk', this is more generalized, and rarely located in a specific place. More critically, the social networks of the pastoralists are highly fluid and open, extending well beyond their local communities. Their closest relatives are often located thousands of miles away, and they invariably have a much larger array of other social and economic relationships that have little or nothing to do with where they live.

Although pioneer history continues to provide a touchstone for many Australians, in specific terms, much of the detailed historical knowledge about the landscape has been lost in the transience of the pastoral population. People coming to work on the cattle stations find that only snippets and faint traces of colonial history remain. Ironically, the older members of the Aboriginal groups previously associated with the station could provide more information, but most of them were pushed off the land and into the Aboriginal reserves when legislation was introduced in the 1960s requiring properties to pay all stockworkers award wages. So the Euro-Australian pastoralists preserve a loose folkloric history of the stations, and can guess at the events which led to names such as Labour in Vain Yard and Battle Creek, but they are mainly reliant upon the sketchy knowledge of previous managers and stockworkers. The provenance of the European place names on the station may be buried in historical archives or explorers' diaries, but for the inhabitants their primary

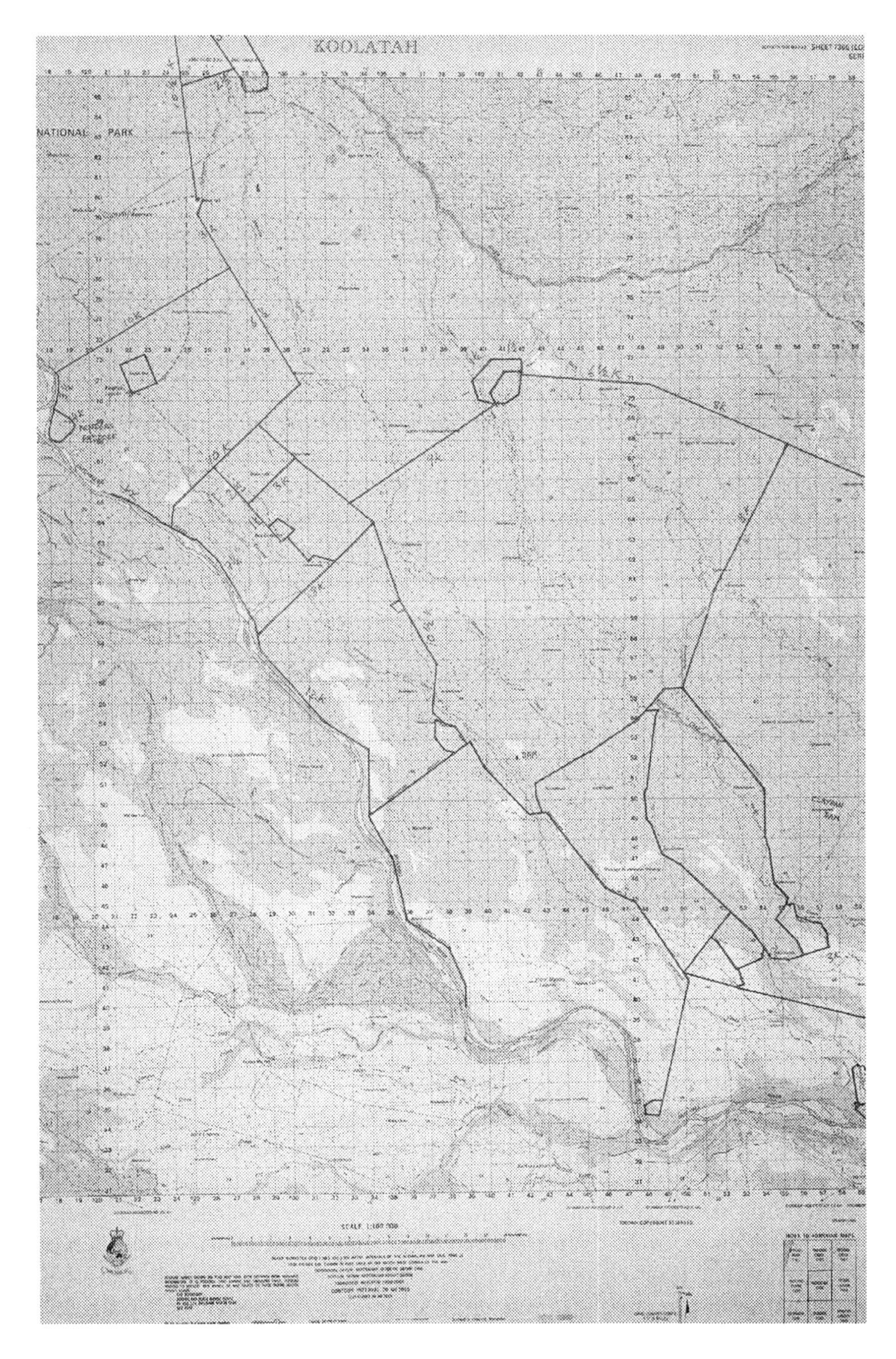

Figure 6.4 Station map of Koolatah, a cattle property near Kowanyama.

meaning is simply that they humanize the landscape in colonial terms. The major working source of information for pastoralists is a standard OS map of the station, upon which fence lines, paddocks, roads and dams have been drawn.

As various writers have noted, (e.g. Cosgrove and Daniels 1988; Bender 1993; Orlove 1991; Morphy 1993), such forms of representation are critical in the construction of knowledge and values relating to place and in defining its socio-political dynamics. These representations depict the landscape in homogenized Cartesian categories, describing its resources generically. Because properties are spatially organised by measured distance and area, the most common names for paddocks and yards are practical ones such as 'ten mile paddock', 'five mile yard', 'six mile dam', which illustrate the dominance of quantitative rather than qualitative terms in visual and oral media.

The pastoralists' representations of the landscape are of course reflective of their primary cosmological model. Based on Western science, this categorizes their environment in largely generic and homogenizing terms. Thus their ecological knowledge is based on universal categories that define all of the aspects of their environment in global rather than local typologies. Areas of land are categorized by the type of timber or grass they have; flora and fauna are defined by their relationship with a global genus. Almost all of these categories are infinitely portable, equally applicable to any and every new environment. The transmission of other information within the cattle station community is also focused on the non-local material of more global knowledges: incoming stockworkers learn the skills of an imported economic mode; children are taught national and international history, geography, language and literature, mathematics and science in a national educational curriculum. Morality and legal principles are broad in scope, reflecting the mores of a much wider society and its relationship with international discourses about human rights and responsibilities.

The pastoralists' economic mode is similarly imposed, having introduced from the other side of the planet technology and animals that bear no relationship to the local ecology. The pastoralists' daily interaction with the local environment is, in reality, more of an interaction with these imposed elements. The stock team spends its time mustering and drafting cattle, breaking horses, and building fences and yards. The domestic members of the cattle station community remain sequestered within the confines of the homestead, cooking for the stock team, laundering and cleaning, looking after children and – often – nurturing many imported plants and vegetables to make green oases in defiance of the bush. These economic activities are not remotely self-sufficient: much food has to be brought from elsewhere, and the station's own production is entirely dependent on economic exchange systems that are national and often global. Thus the pastoralists' economic knowledge, like their social organization, is fully open to a much wider sphere. Their concepts of property are also drawn from a legal system that defines land and resources as alienable commodities. Almost all cattle property is held under leasehold, rather than freehold tenure, underlining the temporary nature of ownership.

The global context of their knowledge systems is reflected in Durkheimian terms (1954 [1912]) in the religious cosmology of the pastoral community. Most are Christians (at least nominally) and so subscribe to spiritual beliefs in which God is located not in the immediate environment, except in terms of a vague and generalized omnipresence, but in a far off heaven. Human spiritual being is similarly amorphous and not connected to place. Even for the most optimistic Christian fundamentalists, of which there are a few in the area, death is seen pretty much as a one way ticket to the stratosphere.

A consistent picture emerges. Like all Euro-Australians (and other Westerners) the pastoralists do have a sphere of local knowledge – as Geertz noted: 'the shapes of knowledge [are] always ineluctably local, indivisible from their instruments and encasements' (1983: 4). However, the fact that practices are situated does not mean that they are necessarily located in place with any permanence. Although the knowledges upon which the pastoralists construct their identity are superficially local, this is only a temporary and partial point of contact that can be readily uprooted and transported elsewhere. Further localization is continually prevented and disturbed by the mobility of the population and the tenuousness of its land tenure. More critically, while areas of knowledge may be applied locally, each is largely dominated by information that is not in fact local in its nature. The pastoralists' cosmos, their social and economic organization, their ecological knowledge, their representations of the land and so forth, all extend well beyond the medium of their local environment and engage with ever widening circles of knowledge which are, eventually, global.

This is not to suggest that the pastoralists have no specific, complex and deeply affective relations with place. Like most people, they are continually attempting to achieve precisely these things, and their rural economy enables them to be more successful in cathecting with place than the vast majority of Euro-Australians. The problem is that they are enmeshed in much wider systems which effectively prevent them from achieving the long-term intimacy of connection, the security of tenure and the certainty of identity to which they aspire. Their knowledges are neither local nor turned inwards to create a distinctive, separate and closed cultural space: they are instead turned outwards, open to the global flow. Equally critically, there is little to integrate the various types of knowledges that they acquire: only the most tenuous relationships between social networks and economic activities, with no obvious and graspable connection between social and spiritual identity. To some degree, this is an issue of scale – essentially, local knowledges can be encompassed and integrated; vast global systems can only be entered partially and dealt with haphazardly – they are beyond the scope of what is cognitively manageable. It is also a practical issue: unlike the indigenous people, the inhabitants of the cattle stations do not use the physical environment as a primary medium for their knowledges, and therefore have no co-ordinating, correlating repository which enables intellectual coherence. It is only in the aspects of their lives that are related to place that their knowledges begin to cohere.

It is no wonder then, that the pastoralists defend their hold on the land with grim determination, and that this perceived stability and connection lies at the heart of Australian identity. It is equally understandable that the majority of the population, continually moving and experiencing the loss of place, looks to its rural heartland and, increasingly, to the Aboriginal community for inspirational visions of permanence, a Nirvana of harmony and belonging.

A matter of life and death

This raises an important underlying issue: in recent years numerous writers (e.g. Foucault 1978; Lowenthal 1991) have elucidated the interrelatedness of knowledge and identity, establishing that, to some extent, knowledge *is* identity – the basis of selfhood. If knowledge is disparate and impossible to integrate, there must surely be an emotional cost. When people are continually required to absorb new places and information, this necessitates the abandonment – at least to some extent – of previous places and persona. Thus Australian pastoralists, like other mobile Westerners, regularly experience the loss of place and selfhood. The intimations of mortality are inescapable: for them, there is no local conception site which contains and replenishes their social being, there is no reunification with a totemic lodestone of ancestral forces, no easy slippage between immediate visible and invisible dimensions, and no implied regeneration. Transcendence of mortality depends on such mechanisms to provide continuity. Among Euro-Australians, famous lives may be carried beyond the grave, celebrated in monuments and records, but for most, for the average cattle farmer in North Queensland, there is little prospect of such perpetuity. The best they can hope for is a scattering of family photographs and heirlooms, maybe a small gravestone in some arbitrary cemetery, and, for a while at least, some fond – or maybe less than fond – recollections by their immediate family and friends.

There are more subtle forms of social death that are also relevant to a discussion about systems of knowledge and identity, and provide some insight into the persistent romanticization of indigenous relations to land. It is evident from the ethnographic data that Aboriginal systems of knowledge, firmly located in place, integrated, and thus provided with massive continuity, are largely shared by all members of the community. Even today, despite the additional information being encompassed, traditional knowledges – 'custom ways' – are still successfully transmitted between generations. Collective knowledge enables forms of social identity that are similarly communal. The inhabitants of Kowanyama have no difficulty in defining who they are: as well as being privy to a readily definable body of knowledge – their ancestral law – they come from a particular place, and are therefore part of a kin network, a clan, a language group and, more recently, a local Aboriginal community.[13]

What happens, though, to the prospect of social being when knowledges are only partially shared, or barely connected in a broad global sphere? Surely this, more than any other factor, is a recipe for individuation and social alienation. There is no global village – this is an oxymoron: a village is human in scale, socially and physically manageable, easily encompassed – it is, in essence, local, and its knowledges are firmly situated in place. For Euro-Australians, the world that is 'the homeland of their thoughts' (Ingold 1995) is too large in scale to allow the knowledges that it contains to be integrated and, accordingly, they demonstrate constructions of identity which are highly individuated and which metamorphose in each new context as they sever ties and move on. This is a subtle social death – a death of a thousand cuts – which is perhaps why the apparent security of indigenous knowledges may seem enviable.

Knowledge and intimacy – reconsidering the romance

In conclusion: the ethnographic comparison outlined above suggests that 'indigenous knowledges' may be defined by a number of common characteristics, all of which are dependent upon and continually reaffirmed by their location in place. Although Aboriginal Australians provide a classic example of a cultural group for whom land is central, there are many other indigenous knowledge systems that may be said to have similar characteristics. Indeed, the question arises as to whether it is even feasible to consider indigenous knowledge that is not, in some way, held in place.

If the defining characteristic (under which all others are subsumed) is that indigenous knowledge and identity is place-based, this supports the argument that it is indeed significantly different from other forms of knowledge in that its particular characteristics cannot be replicated on a larger, global scale. Also implicit in this comparison, and tucked into the sub-texts of debates on indigenous knowledge, is the even more contentious idea that some ways of creating knowledge and identity and interacting with the environment may be more 'natural' or fulfilling than others. As Milton points out, 'emotions are fundamental to the process of learning' (2002: 148). It is difficult to discuss what may be 'natural' without backsliding into essentialist notions of 'human nature'. However, there are surely some legitimate and useful questions about evolutionary adaptations and the kinds of knowledge systems that human cognitive processes are best suited to construct. The outpouring of angst in discourses about globalization suggests that this is a process that generates considerable anxiety. How much of this anxiety, and the yearning for the perceived characteristics of indigenous models, is related to the homelessness and alienation noted by Berger (Berger, Berger and Kellner 1973)? Do Westerners, adrift in vast systems of knowledge, feel that they have indeed lost their place?

Modern anthropologists have gone to some lengths to demonstrate that environmental relationships are essentially social, and to integrate discourses on social and environmental issues. Some (e.g. Milton 1993, 1996; Ellen and

Fukui 1996) have made valuable critiques of the discourse of the environmental movement, pointing out that its specialized focus on nature disconnects ecology from society, with massive implications for developmental efforts in many parts of the world. Descola and Palsson (1996) have argued that this depends on a putative dualism between nature and culture which does not acknowledge that they are mutually constitutive. As Seeland puts it: 'In the Western world, nature is commonly perceived as separate from human culture and civilization' (1997: 1). Dwyer suggests that this is because 'Western thought has mistaken the periphery for the primal. To a large extent the conservation movement has compounded the error by sanctifying the perceived primal. The consequence is alienation' (in Ellen and Fukui 1996: 179).

However, in affirming the socio-cultural basis of environmental relationships, little has been made of the relationship between global knowledges and social alienation, and the aspirations for community that are contained in alternative representations of indigenous groups. In deconstructing the illusory images of the green movement (e.g. Ellen 1986), it may be that anthropologists have failed to acknowledge the importance of ecology as a metaphor for social being. Yet it is clear that when environmentalists hold up indigenous groups as an example of harmony with nature they are not merely talking about some kind of imagined ecological balance, they are pointing to ways of being in the world that are perceived as more socially and emotionally coherent. It is reasonable to argue that they are, in fact, articulating a yearning for a more stable society in which knowledge is local and integrated, offering a manageable whole, a firm connection with place, and a sense of belonging. Underpinning this model is the assumption that this cathection will generate a commensurately deep concern for the ecological well-being of the environment, but to present this as the major goal is perhaps to miss the point.[14] As Douglas and Wildavsky pointed out, there are – invariably – implicit social agenda in alternative representations:

> Although Friends of the Earth rejects the idea that its own organization may serve as a model for future society, its long-term vision is *a harmonious world* [my emphasis] modelled on its concept of balanced interrelations with nature . . . (Douglas and Wildavsky 1982: 137)

Though it is doubtless valuable to encourage the rejection of romantic stereotypes, it may be that anthropologists have been a little too careful to distance themselves from any association with such reifications of indigenous knowledge. It is worth bearing in mind that, in the political environment in which this discourse is conducted, 'greenies' and 'new age' groups who oppose capitalist ideology and the hegemony of globalization in its various forms are continually subject to a heavy media assault from interest groups who hope to gain greater access to land and resources. No discipline is entirely immune from these influences, and anthropologists should perhaps be wary of being suborned by avid debunkers with an interest in homogenizing knowledge and dismissing or suppressing alternative models. In Australia, for example, such absolute cultural relativity would leave indigenous groups with little defence against further colonial appropriation. There are already worrying shifts in precisely this direction, as a powerful heritage movement strives to establish the idea that Euro-Australian connection to place is not substantially different to that of the indigenous population. The implications of these efforts for the land rights issue are severe.

Similarly, if indigenous knowledge is defined in purely ecological terms that can be conveniently incorporated into Western scientific models, the integration of local ecology with social and spiritual life that lies at the heart of emic definitions of knowledge will be denied. This is a problem that dogs development endeavours, and has often produced inappropriate programmes unsympathetic to indigenous views. It is perhaps the task of anthropologists to point out that metaphorical languages, such as the ancestral mythology of Aboriginal Australians, are as effective in describing a social and physical landscape as any literal Cartesian terminology, and indeed may be better suited to expressing the complexities of human-environmental relationships.

Geertz (1983: 57) exhorted anthropologists not to be 'systematically deaf to the distinctive tonalities' of indigenous knowledges. Perhaps it is time to consider that these distinctive tonalities may be significantly different from those in globalized systems of knowledge. The ethnographic evidence presented above suggests that the local knowledge does offer more than the sum of its parts. It attempts to demystify this difference by examining the everyday practices through which knowledge and identity are localized: for example, the use of the land as a mnemonic and as a medium for integrating knowledge; the associated qualitative and metaphorical discourses; the limitations of scale in indigenous socio-economic forms.

The ethnography outlined here suggests that, compared with environmental relationships which deny the localisation of knowledge, indigenous knowledge systems may offer greater potential intellectual and emotional stability, psychological support in confronting mortality, and real potential for cathection with place. As such, they appear to meet the aspirations of many Westerners for precisely these qualities, and it is thus entirely logical that they should be held up as a model by organizations concerned with these issues. Implicit in this comparison is a critique of knowledges that are fluid, large-scale, specialized and fragmented, which opens up some useful questions about the social and psychological effects of discontinuity and displacement. It seems, therefore, that indigenous knowledges have much more to offer anthropologists than mere ecological details that can be appended to the categories of Western science.

Notes

1 This assumes that 'knowledge' may be defined in any area of cultural life, rather than according to a narrower Western definition of 'scientific' knowledge.

2 In Kowanyama, the common term for a totemic ancestor is *ping a nim,* which may be translated as 'mate' or 'companion'.

3 The name for an individual's spiritual conception site is revealing, pointing to the fact that in Aboriginal terms the Dreaming is more of a place than a time. Thus in *errk*

elampungk, *errk* is 'place' *el* is 'eye' and *ampungk* is 'home', thus it can be translated as 'the home place of your image'.

4 A clear indication of the concept that each generation is reincarnated is provided by kin terminology in which grandchildren are called by terms for previous generations: 'It goes down from one generation to another, if she – my grand daughter – have a baby, I have to call that baby little auntie or uncle/father' (Alma Wason).

5 Although the bringing together of different language groups in mission communities such as Kowanyama has led to many 'wrong way' marriages, and the rules have become more relaxed in recent years, there is still a clear idea of 'proper' or 'right way' marriage based on Ancestral Law.

6 Ego will marry the DD of a woman who is his classificatory FM.

7 It is not feasible to draw clear boundaries between Aboriginal language groups, particularly in densely populated regions such as Cape York. Prior to colonial dispossession, it seems that most Aboriginal people spoke several languages, and that there were many small linguistic sub-groups. Thus a particular language might be spoken predominantly in one area, but at its margins it would be common to find dialects and use of neighbouring languages, which supports the evidence for a network of exchanges over considerable distances.

8 This can be described as an intense psychological bond – a deeply inculcated form of connection.

9 Aboriginal people in Kowanyama, like those in most parts of Australia, have a complex series of rights to land, based on kin relationships. People's primary land ownership (in this region) is generally of their father's country, in keeping with the system of patrilineal descent, but they will have major secondary rights to their mother's country, and rights of usufruct in the clan country of other kinfolk, based on their relative geographic and genealogical closeness.

10 This localization of knowledge is of course not confined to hunter-gatherers. Many nomadic pastoral peoples, who may travel vast distances, are reliant upon a series of specific, known locations that can be relied upon to provide resources for an annual economic cycle.

11 Even where people have been alienated from their own land, they continue to use representations of pre-colonial landscapes as a basis for identity. This is enabled by massive historical momentum, but there are some difficult questions about the long-term feasibility of such a practice in the absence of ownership of the land and every-day interaction with it.

12 Writing about peasant societies (1991), Gellner argued that people draw their identity from the land in proportion to their level of involvement with it.

13 In Aboriginal communities such as Kowanyama there is now a clear sense of pan-Aboriginality, but although this is politically important, most particularly when the community engages with the legal issues relating to land rights, it remains, like the overlay of Christianity, merely an addition to the more focused realities of daily life.

14 It is interesting to note the preponderance of communitarian terms in the names of so-called environmental organizations: *Friends* of the Earth, Green*peace,* Wildlife *Trust.*

References

Altman, J.C. 1984. Hunter-gatherer subsistence production in Arnhem Land: The original affluent society hypothesis re-examined'. *Mankind* 14 (3): 179–190.

Bender, B. (ed.). 1993. *Landscape, Politics and Perspectives.* Oxford, New York: Berg.

Berger, P., B. Berger and H. Kellner. 1973. *The Homeless Mind.* New York: Vintage.

Bourdieu, P. 1977. *Outline of a Theory of Practice.* Transl. R. Nice. Cambridge: Cambridge University Press.

Charlesworth, M., H. Morphy, D. Bell and K. Maddock (eds). 1984. *Religion in Aboriginal Australia.* Queensland, Australia: University of Queensland Press.

Cosgrove D. and S. Daniels (eds). 1988. *The Iconography of Landscape.* Cambridge: Cambridge University Press.

Descola, P. and G. Palsson (eds). 1996. *Nature and Society: Anthropological perspectives.* London, New York: Routledge.

Douglas, M. and A. Wildavsky. 1982. *Risk and Culture.* Berkeley, Los Angeles, London: University of California Press.

Durkheim, E. 1954 [1912]. *The Elementary Forms of the Religious Life.* London: Allen and Unwin.

Dwyer, P. 1996. The invention of nature. In *Redefining Nature: Ecology, culture and domestication.* (eds) R. Ellen and K. Fukui. Oxford: Berg.

Ellen, R. 1986. What Black Elk left unsaid: On the illusory images of green primitivism. In *Anthropology Today,* 2 (6): 8–12.

Ellen, R. and K. Fukui. 1996. *Redefining Nature: Ecology, culture and domestication.* Oxford: Berg.

Foucault, M. 1978. *The History of Sexuality.* Vol 1. Harmondsworth: Penguin.

Geertz, C. 1983. *Local Knowledge: Further essays in interpretive anthropology.* New York: Basic Books.

Gellner, E. 1991. Le nationalism en Apesanteur. *Terrain* Vol. 17: 7–16.

Gosden, C. 1994. *Social Being and Time.* Oxford: Blackwells.

Gould, S. 1987. *Time's Arrow, Time's Cycle: Myth and metaphor in the discovery of geological time.* London: Penguin Books.

Gurung, B. 1997. The perceived environment as a system of knowledge and meaning: A study of the Mewahang Rai of eastern Nepal. In *Nature is Culture: Indigenous knowledge and socio-cultural aspects of trees and forests in non-European cultures.* (ed.) K. Seeland. London: Intermediate Technology Publications Ltd. 19–27.

Hamilton, A. 1982. Descended from father, belonging to country. In *Politics and History in Band Societies.* (eds) E. Leacock and R. Lee. Cambridge: Cambridge University Press.

Hirsch, E. and M. O'Hanlon (eds). 1995. *The Anthropology of Landscape: Perspectives on place and space.* Oxford: Clarendon Press.

Ingold, T. 1995. Building, dwelling, living: How animals and people make themselves at home in the world. In *Shifting Contexts: Transformations in anthropological knowledge.* (ed.) M. Strathern. London, New York: Routledge. 57–80.

—— *The Perception of the Environment: essays on livelihood, dwelling and skill.* London and New York: Routledge.

Ingold, T, D. Riches and J. Woodburn (eds). 1988. *Hunters and Gatherers 1: History, evolution and social change.* Oxford, New York, Hamburg: Berg.

Jones, R. 1990. Hunters of the dreaming: Some ideational, economic and ecological parameters of the Australian Aboriginal productive system. In *Pacific*

Production Systems: Approaches to economic prehistory. (eds) D. Yen and J. Mummery. Canberra: Australian National University. 25–53.

Layton, R. 1989. *Uluru: An Aboriginal history of Ayers Rock.* Canberra: Aboriginal Studies Press.

Levi-Strauss, C. 1968. *The Savage Mind.* USA: University of Chicago Press.

Lowenthal, D. 1991. British national identity and the English landscape. *Rural History* 2 (2): 205–230.

Milton, K. (ed.) 1993. *Environmentalism: The view from anthropology.* London and New York: Routledge.

—— 1996. *Environmentalism and Cultural Theory: Exploring the role of anthropology in environmental discourse.* London: Routledge.

—— 2002. *Loving nature: Towards an ecology of emotion.* London, New York: Routledge.

Morphy, H. 1988. Maintaining cosmic unity: ideology and the reproduction of Yolngu clans. In *Hunters and Gatherers 2. Property, power and ideology.* (eds) T. Ingold, J. Riches and J. Woodburn. Oxford, New York, Hamburg: Berg.

—— 1991. *Ancestral Connections: Art and an Aboriginal system of knowledge.* Chicago: Chicago University Press.

—— 1993. Colonialism, history and the construction of place: The politics of landscape in Northern Australia. In *Landscape, Politics and Perspectives.* (ed.) B. Bender. London, Providence: Berg. 205–243.

—— 1998. Cultural aspects of adaptation. In *Human Adaptation.* (eds.) G. Harrison and H. Morphy, Oxford, New York: Berg.

Morton, J. 1987. The effectiveness of totemism: 'Increase ritual' and resource control in central Australia. *Man.* 22: 453–474.

Munn, N. 1973. The spatial presentation of cosmic order in Walbiri iconography. In *Primitive Art and Society.* (ed.) A. Forge. Oxford: Oxford University Press.

—— 1984. The transformation of subjects into objects in Walbiri and Pitjantjatjara myth. In *Australian Aboriginal Anthropology.* (ed.) R.M. Berndt. Nedlands: University of Western Australia Press.

—— 1986. *Fame in Gawa: A symbolic study of value transformation in a Massim Papua New Guinea society.* Cambridge: Cambridge University Press.

Myers, F. 1986. *Pintubi Country, Pintubi Self: Sentiment, place and politics among Western Desert Aborigines.* Washington: Smithsonian Institute Press.

Orlove, B. 1991. Mapping reeds and reading maps: The politics of representation in Lake Titicaca. *American Anthropologist.* 3–40.

Read, P. 1996. *Returning to Nothing: The meaning of lost places.* Cambridge, New York, Melbourne: Cambridge University Press.

Rudder, J. 1983. *Qualitative Thinking: An examination of the classificatory systems, evaluative systems and cognitive structures of the Yolngu people of northeast Arnhem Land.* Unpublished MPhil thesis, Australian National University.

Seeland, K. (ed.). 1997. *Nature is Culture: Indigenous knowledge and socio-cultural aspects of trees and forests in non-European cultures.* London: Intermediate Technology Publications Ltd.

Sharp, L. 1937. *The Social Anthropology of a Totemic System of North Queensland.* Unpublished PhD thesis. Harvard University.

—— 1939. Tribes and totemism in north-east Australia. *Oceania* 9 (3): 254–275

—— 1952. Steel axes for stone age Australians. In *Human problems in technological change: A casebook.* (ed.) E. Spicer. New York: The Russell Sage Foundation.

Strang, V. 1997. *Uncommon Ground: Cultural landscapes and environmental values.* Oxford, New York: Berg.

—— 1998. The strong arm of the law: Aboriginal rangers and anthropology. *Australian Archaeology* 47: 20–29.

—— 1999. Familiar Forms: Homologues, culture and gender in northern Australia. *Journal of the Royal Anthropological Society* (N.S.) 5: 75–95.

Street, B. 1975. *The Savage in Literature: Representations of 'primitive' society in English fiction 1858–1920.* London, Boston: Routledge.

Sutton, P. 1978. Aboriginal Society, Territory and Language at Cape Keerweer, Cape York Peninsula, Australia. Unpublished PhD thesis, University of Queensland.

Taylor, J. 1974. *Of Acts and Axes: An ethnography of socio-cultural change in an Aboriginal community, Cape York Peninsula.* Unpublished PhD thesis, James Cook University, Queensland.

Tilley, C. 1994. *A Phenomenology of Landscape: Places, paths and monuments.* Oxford, Providence: Berg.

Williams, N. and E. Hunn (eds). 1986. *Resource Managers: North American and Australian hunter-gatherers.* Canberra: Australian Institute of Aboriginal Studies.

Wilson, P. 1988. *The Domestication of the Human Species.* New Haven: Yale University Press.

Chapter 7

International animation

UNESCO, biodiversity and sacred sites

Terence Hay-Edie

As a grand ethical endeavour, the founding constitution of the United Nations Educational, Scientific, and Cultural Organization (UNESCO) states, 'that since wars begin in the minds of men, it is in the minds of men that the defences of peace need to be constructed'. From its inception, UNESCO's mission statement thus positions the organization as a pre-eminently intellectual pursuit. However, quite unlike related academic disciplines, the UN specialized agency remains an international development agency officially composed of governments. Nation states, which constitute UNESCO's core membership, provide the institution with regular budgetary contributions and expect it to operate as an efficient bureaucratic body capable of administering a range of specialized programmes.[1]

This anthropologist's interest in UNESCO stems from a one-year engagement within the division of ecological sciences at the organization's headquarters in Paris between 1995–1996. In the description that follows, I go on to briefly introduce my own involvement as anthropologist in the conceptualization of an inter-sectorial programme designed to revalorize forms of 'vernacular conservation' of biodiversity based on indigenous knowledge. The proposed initiative provided a useful point of departure to explore a network of relations beyond the 'tower of glass' of the UNESCO-HQ secretariat, including ethnographic research at the annual Working Group on Indigenous Populations held at the United Nations in Geneva, as well as during a regional seminar held in India.

With over 30 regional offices worldwide, UNESCO is both scattered geographically, as well as dispersed conceptually in an array of symbolic transformations. Fieldwork thus attempted to engage multiple facets of the organization at headquarters, as well as in associated events further afield. By tracking a range of different UNESCO activities, I encountered numerous other actors also copying and adjusting themselves to fit the broad institutional discourse of the international organization. Much of the UNESCO process emanating from the central headquarters thus revolved around a template of action offering predetermined global categories, unfinished with local details, waiting to be sculpted or filled in at the periphery.

Defining the sacred in UNESCO

Since 1996, staff from different divisions in UNESCO had begun to assemble documents and information on how landscape features treated as sacred by local or indigenous people could successfully conserve biological diversity. As a contribution to the aims of ethnobiology to revalorize forms of traditional ecological knowledge (TEK), it was pointed out that the role of 'natural sacred sites' was attracting increasing interest from the World Wide Fund for Nature,[2] the Mountain Institute, and had significant relevance for the implementation of the Convention on Biological Diversity (CBD) drawn up at the Rio Earth Summit. Questions were thus raised concerning the position of sacred sites as 'vernacular protected areas' and their relation to other profane areas, and how to scale-up from individual cases to a broader comparative level. Some of the following working hypotheses were suggested:

1 Do sacred sites come to have similar sanctions on cutting wood or the collection of certain plants; as the location of tombs or ancestral shrines managed by sacred specialists; as loci for finding medicinal plants (Schaaf 1995)?
2 Do sacred sites occur equally in both resource-rich and resource-poor ecosystems and can the gene-pools of plants in sacred sites be used in programmes of restoration ecology (Ramakrishna 1996)?
3 In what measure are these sacred sites natural? Does one find in them vestiges of primary forest or, rather, anthropogenic habitats where certain plants are encouraged or planted as agroforests to provide communities with selected natural resources? Can these sites therefore be examined as indicator sites for assessing the potential natural vegetation of ecosystems degraded or modified by human impact (Roussel 1992)?
4 What is the overlap between cultural and spiritual values of biodiversity and scientific or economic valuation techniques?

One possible interpretation of the proposal, associated with strong critics of the development encounter such as Escobar (1994), might be to cast the programme as fundamentally hegemonic. Reluctant to overtly impose its own brand of nature protection, was UNESCO in the business of appropriating cultural idioms instead? A more sympathetic view, resulting from a prolonged association with proponents of the concept, arises from staff who posed themselves the basic question of 'whose reality counts' towards the conservation of biodiversity. In fact, as a number of the civil servants admitted, the innovative proposal entitled 'Sacred Sites – cultural integrity and biological diversity' actually ran against the grain of most single disciplinary-based donor bodies, limiting its chances of attracting external funding.

Nonetheless, for purposes internal to UNESCO's own desire to harmonize its different sectors, the protection of sacred sites was deemed to fall squarely within the fields of both Culture and Science. Amongst its many publications, the Culture sector had, for example, produced a manual entitled 'The Cultural Dimension of Development – towards a practical approach' citing numerous ineffective development ventures. In one instance, the book referred to a joint project by the World Bank and the Ecuadorian Ministry of Agriculture which had failed to boost the productivity of guinea-pigs reared in the kitchens of Andean homes as a vital source of protein, concluding 'from the very beginning, therefore, an anthropologist has to be called upon to make a study of the whole cultural environment of the project' (1995: 19).

From a cultural standpoint, sacred sites could in this manner also reflect unique associations with the environment linked to the lifestyles of local or indigenous cultures close to the land.[3] Through a scientific lens, many of the sacred groves found in parts of East and West Africa, South Asia and the Pacific, could potentially also represent, as one staff member put it, important indicator sites for assessing potential natural vegetation. Despite these common interests, however, protagonists from the Culture sector continually emphasized a focus on 'processes' over the course of brainstorming meetings, while those from the Science sector often stressed the spatial identification of 'sites' and their overlap with internationally-recognized protected areas such as biosphere reserves and world heritage sites.

Located within the Science sector, I was also able to participate in the debate suggesting that sacred sites could only assume an identity by the delineation of boundaries with distinct markers and access restrictions such as found in the relict Kaya forests of coastal Kenya (see Parkin 1991; Mutoro 1994). As a fieldwork technique based in a self-referential institution, the researcher could thus contribute a critical perspective from within the discipline of social anthropology, while also deploying it as a 'node of enunciation' within a conceptual network. Yet at times, this meant using radically different interpretations of the same term. The hold-all concept of a sacred site acted as a sort of 'trap' (Gell 1999) and increasingly became the object of my analytical interest in the subsequent examination of its retranslation.

However, as I negotiated the fine line between applied hubris and academic tentativeness, I often felt precarious in my role of participant comprehension in the proposed programme. At one point, while editing an information circular, I was accused by another French social anthropologist as '*vraiment new age*'.[4] Yet, from the vantage point of participatory involvement in the programme, it became evident that staff within UNESCO were often determined to avoid any false romanticization of an 'ecologically noble savage' (cf. Ellen 1986) or to present crude depictions of 'traditional' societies.

After one particular meeting, a member of the Culture sector invited me to his office to discuss the concept for the programme. High in the UNESCO building with a view over the roofs of Paris, he pointed out that the 'ability to summarize comes in useful in UNESCO'. Mentioning another workshop he was organizing on the theme of biocultural diversity in one of the major regional offices, he added that the organization was ill-equipped to deal with the topic of the sacred. There could be no answer, he continued, as to how the institution apprehends the sacred as 'it is impossible for the intellect to grasp the sacred'. He felt, nonetheless, that the discussions had made substantial progress on reaching a level of 'emotional integrity' towards the institution's own activities.

Sitting in the seemingly prosaic setting of an office with the characteristic filing cabinets, 'in' and 'out' trays, he went on that the search for the sacred was the 'search for truth for the whole of the macrocosmos'. Yet much of the everyday work in the bureaucracy, he continued, related to international conventions and policy statements marked by a constant need to defuse the political implications of 'Culture', as opposed to an ability to see projects 'on the ground' through to fruition. Perhaps paradoxically, he also observed, where UNESCO had engaged in cultural promotion work around the world, the 'symbolism of the act' seemed to enhance the self-image of the people in question, representing to him a form of the 'blessing of UNESCO'.

Exploring the network

In order to investigate relations outside of the UNESCO secretariat itself, I participated in the annual United Nations Working Group on Indigenous Populations (WGIP) held in Geneva in July 1997. As an attempt to chart a series of connections, I hoped to explore the linkage of the event with UNESCO's work in the same field, and to trace the experience of some of the indigenous delegates who attended the event. Since its creation some 15 years earlier, increasing numbers of such representatives from different regions (including South Asia, East Asia, Africa and the former USSR) had been gathering at the WGIP during the 1990s to assert that they had suffered a similar fate from colonial and state persecution, and to try to seek some redress from the UN.

Quite dazzled by the impressive venue, many of the recent arrivals seemed to place considerable faith in the efficacy of simply being heard in the illustrious setting. Nirmal Rai, one of the two delegates from Nepal, had been fortunate to be able to participate by limited travel funds from the UN Sub-Commission on Human Rights. A school teacher who spent at least half the year away from Kathmandu, he was amongst many of the new recruits who had little to no knowledge of the confusing institutional processes compared to other veteran campaigning groups from Australia, the United States, Canada and South America. Still extremely inexperienced in the protocol of the event, Rai signed up to take the floor on the first day under the wrong session and was eventually forced by the conference chair to give up his presentation.

Over a week of recorded interviews and conversations, it became apparent that the main purpose of the ritual in fact hinged on a poignant sense of solidarity between the disparate groups who stressed in unison that 'the land is our mother', and that the draft declaration on their cultural rights should be adopted by governments *in toto* or not at all. Siebert (1996), who also conducted fieldwork at the WGIP, echoes this view observing that the regular prestation rests on an affirmation of mutual inter-definition by the indigenous actors themselves in an *esprit de corps.* As the only such forum within the multilateral system of sovereign nation states, the venue continues to offer some scope to re-negotiate the semantic definition of 'peoples' and 'nations' referred to in the founding charter of the transcendental body. As a Theravadin Buddhist monk from Bangladesh, conspicuous in his bright saffron robes, commented to me after speaking to a television company, 'the UN is the ultimate salvation'.

Set in a palatial building surrounded by sumptuous gardens, the WGIP thus provided an opportunity for dramatic displays of indigenous identity including colourful costumes, music, singing and prayers.[5] Yet the dramaturgical dimension of the occasion, also raised a number of questions regarding how the event translates (if at all) into effects on the ground, as well as in the programmes of the other international organizations such as UNESCO present as observers. As ethnographer, how could one make sense of the evanescent phenomenon as it appeared only once a year for five days with the component influences fragmenting across the world? Did some of the pieces re-assemble into smaller subgroupings in different regions?

I could find few precedents of anthropological investigation. Little's attempt to delimit a 'mega-event ethnography' at the UNCED Earth Summit held in Rio de Janeiro in 1992 portrays anthropological enquiry as a form of privileged journalism recording ostentatious protocol as mere 'political cosmology'. Arguing that the dramatic actualization of the mega-event could be considered, following Tambiah's dichotomy, a type of 'performative ritual', Little felt that 'the information content of the speeches was not of crucial importance', comparing the meeting to a 'speech marathon' (1995: 275). In such a cacophony of voices, he also feared the contribution of anthropology would be difficult to detect:

> The Rio Conference had over 9000 officially registered journalists while only a handful of anthropologists were present at the event as observers. With so many journalists 'creating texts' about the event, is there any room for texts coming from another trade? Does the discipline of anthropology have something different to offer? This competition was not present when ethnographers wrote about the Kwakiutl potlatch or the Trobriand kula.
>
> (ibid: 282–283)

UN mega-events do indeed feature a high degree of repetition of key words, conventionalized etiquette, and a defined aesthetic of document production. Like the UNCED conference, the WGIP was characterized by content overload and apparent redundancy in speech-giving, resulting in low attendance (and attention) levels in the main plenary hall by delegates after the opening ceremony. But what of the wider resonance of such mega-prestations marked by a series of ripples emanating from the initial fanfare, or those of regular fixtures such as the WGIP with a distinct history and evolution? As Little himself remarks, the transient Rio meeting could not be housed in a neat frame as it was the output of months of preparatory work, and would act as the 'symbolic referent for all future conferences' in different locations for years to come.

Ethnographically, these linkages and references seemed a greater challenge, guiding my eventual research concerning the notion of the sacred across different institutional contexts. The fact that self-styled indigenous representatives were invoking the sacred suggested an attempt to place certain areas of discourse outside the realm of a mundane and negotiable sphere of relations, a concern also shared by protagonists in the UNESCO sacred sites initiative. Wright provides a similar description of a UNESCO International Conference on Cultural Policies for Development, describing how the ability to re-cast or stick new meanings onto global categories was located in the real activity of the conference [which] is the corridor conversations and networking' (1998: 176). As Wright asks of the meeting, 'how were some participants trying to denaturalize dominant ideas while others tried to make alternative ideas "unimaginable" or "unspeakable"?' (ibid: 175).

During the WGIP, however, UNESCO was seldom in evidence except during a parallel workshop where a staff member from the Culture sector in Paris presented the findings of the World Commission on Culture and Development, 'Our Creative Diversity', which as she explained, had an implicit bearing on the concerns of indigenous people (see Perez de Cuéllar 1995). In theory, there was a connection to be established between UNESCO and the WGIP, but, in practice, experienced indigenous delegates were focusing their attention elsewhere on Article 8j of the (legally-binding) Convention on Biological Diversity.[6]

In a discussion with the key professional in the UN Commission on Human Rights responsible for indigenous issues, he commented to me that there was indeed significant mutual interest between UNESCO's work and the WGIP, but the establishment of a strong linkage had never got very far. If possible, he hoped to arrange for the WGIP to be convened one year in Paris with the support of UNESCO staff he had met. The connection between the two institutions, although not without great potential, was weak and needed animation. Nonetheless, the world heritage 'cultural landscape' category and the possibility of a programme on sacred sites offered considerable scope for overlap, but the association had yet to be firmly established.

Having pursued a potential avenue of research thus far, in an attempt to triangulate some of the different spheres of relations within UNESCO and the concerns of indigenous peoples at the WGIP, the existing density of connections, although latent, did not appear significant enough to warrant the status of a delimited network. Multiple institutions and interests had coalesced, if only briefly, while at the same time maintaining an internal focus on linkages within their own separate but distinct areas of activity.

Promoting the UNESCO sacred sites initiative

In order to launch the above-mentioned initiative on sacred sites, UNESCO-HQ had also been soliciting allies from within its own network of regional offices and national commissions. Following a communication circular regarding the proposal, the UNESCO office in New Delhi was one of the first to respond positively by mobilizing funds for a 'Regional seminar on the role of Sacred Groves for the Conservation and Management of Biological Diversity' to take place at the Kerala Forestry Research Institute (KFRI) in India in December 1997. To this end, in consultation with a member of the division of ecological sciences, Max Ecoworthy, I helped prepare a comparative framework for sacred sites. As a joint article, the piece hybridized an institutional advertisement for UNESCO, an anthropological argument concerning memory and landscape, as well as a reflexive call for criticism and feedback from the seminar. As Ecoworthy annotated the text:

> The stimulus for this initiative comes from several different programmes and interests, including concern with traditional ecological knowledge, so-called 'vernacular conservation' and bottom-up approaches to conservation, the notion of 'cultural landscapes', and cultural dimensions of natural resource use. Questions raised concern the impact of existing or potential [UNESCO] Biosphere Reserve or World Heritage 'cultural landscape' status on local sacred sites. Can international protected areas serve to dislocate local peoples from their symbolic ties to land, or are local systems of values recognised and effectively protected? A number of field projects have already encompassed aspects of the 'sacredness-culture-biodiversity' triptych as an ensemble addressing a wide range of the perspectives.

Shortly before the meeting was due to take place, however, Ecoworthy withdrew his participation in the seminar and passed the responsibility to another colleague, Magnus Theomann, to deliver the UNESCO institutional address and oversee the consultation on possible future case studies. As a consequence, this also left me with the task of delivering the thematic paper. On the first day, after a candle-lighting ceremony and Vedic hymn announced as the 'wisdom of the ancients', the director of the UNESCO office in India stated proudly during his opening remarks that 'we have now learnt to cope with many realities'. He asked whether the audience had heard the 'tale of the British anthropologist' who on witnessing an old man dancing in the forest had mistakenly thought him to be alone while he was, in fact, 'really with the nature spirits'? Shifting his stance towards the discipline, he went on to emphasize the 'vital role of the 1972 World Heritage Convention which now encompasses a more anthropological vision of the world [since it] adopted in 1992 the cultural landscape category'.

The director then explained that the participants were honoured to have a representative from UNESCO in Paris who had travelled a 'long, long way' to take part. Theomann then took the podium for his keynote address where he affirmed the principle of a 'stewardship approach' to conservation. For this to work effectively, he asserted, 'communication' required that the 'different intellectual and analytical language' of scientists and government officials, if combined with the 'mystical aspects of local knowledge' would need 'careful horizontal coordination for the custodians to be considered as eco-partners'. How could these 'institutional barriers' be overcome he wondered? Responding to his own question, Theomann reiterated UNESCO's celebrated capacity for interdisciplinarity stating that:

> UNESCO is probably the best placed organization to deal with such a topic: with our mandates in science (environmental sciences) and culture we want to bridge the sometime artificial gap between nature and culture. We have had a lot of feedback and positive reactions from such diverse countries as Canada, China, Germany, Kenya, Madagascar and Mexico – as well as countries represented at this workshop. International and national organizations such as FAO, IUCN, IDRC, the Mountain Institute and the German Federation for the Protection of Nature want to participate in this initiative with UNESCO.[7]

Following Theomann, the convenor of the seminar, Murtiraja, a professor from New Delhi with years of contact with UNESCO-HQ took the floor. After prolonged ecological research in northeastern India on shifting cultivation and soil nutrient recycling, Murtiraja had witnessed many areas where the only healthy forests left standing were sacred groves, persuading him of the importance of cultural factors in conservation. More recently, he had also carried out a survey in the foothills of Mount Kanchenjunga in the state of Sikkim, which had converted him to the notion of the sacred landscape as a means of integrated conservation combining a variety of altitudes and ecological zones into a single protected area. Having spent much time participating in the International Human Dimensions Programme (IHDP),[8] Murtiraja switched with apparent ease between statements regarding Indian philosophy to those of ecological planetary prophecy. In a presentation entitled Conserving the Sacred, he thus extended his argument to cover:

> The exaltation of the World as the body of God as enshrined in the ancient Indian tradition – the Vedas of antiquity . . . through the sacred landscape linked through the Ganga river system right from the highest mountain reaches of the Goumukh in the Garhwal Himalaya extending up to the Gangetic delta merging into the bay of Bengal, a few hundred kilometres away. Tat Twam Asi, literally translated is 'That Thou Art', is a very profound Vedantic statement, from the Hindu scriptures of antiquity. Here the individual (Twam) is identified as part of the creation – the Brahman

> (Tat) ... in an ecological sense [it] projects the individual as part of Prakriti or the world as we perceive it, the 'Nature'... This would then imply compassion and fellowship, not just between humans but extended to the entire biosphere. Such a relationship alone will give meaning and value to our world, lest we fall into the kind of anthro-centred thinking that has been the bane of traditional ecological thought.

This dual form of globalism could, on the one hand, be attributed to the considerable scope of Hindu thinking, in other words uniquely Indian or South Asian; on the other, to Murtiraja's nodal position in a contemporary international network on the health of the biosphere.[9] In contrast, a social anthropologist long accustomed to analysing such old chestnuts as the nature/culture divide, Douglas (1987) explores the use of the term 'world' as applicable to distinguishable 'theology worlds, anthropology worlds and science worlds'.[10] In an apt demonstration of the notion of such worlds at an ambitious scale, Murtiraja thus moved interchangeably between international UNESCO/IHDP 'biosphere thinking', or could alternatively reconfigure the planet according to ancient Indian scripture.

On the whole, the seminar revolved around two major themes: first, given that sacred forests in India are generally less than one hectare in size, the search for an adequate definition for what could be considered a grove; and second, concerning how the famous (Durkheimian) dichotomy between the sacred and the profane could or could not be applied to concepts of nature in South Asia. During his opening speech, Theomann had claimed that sacred areas could be 'reservoirs of biological diversity' separated from 'neighbouring land which is profane', citing a linkage between the chieftanship of Ghanaian village shrines and the potency of sacred groves 'relying on animist beliefs'. Like other participants, Murtiraja cautioned however that there might be a tendency 'to identify a boundary according to what we are looking for'. Yet, he also felt that the grove unit was a 'constant point' between the concepts of 'single sacred species and sacred landscapes with high levels of biophysical connectivity'. How wide, Theomann then wondered, could the 'catchment area' of such a sacred site actually be:

> The physical area of a sacred site may vary from a single object of veneration to a whole landscape, but also the dwelling area of people who venerate the sacred site may range from local to global. While the Kaaba in Mecca or St. Paul's Cathedral in Rome have a global outreach and significance for Muslims and Catholics respectively, a shrine in an African village may have a local significance for the specific village community only. It may be interesting to discuss at this workshop also the 'area of influence' of a sacred grove or its outreach as this would probably have a major importance on the long-term and spatial protection status of the sacred site.

As the seminar developed, a number of regional examples were suggested. Participants invoked a range of tribal groups for their capacity to protect groves such as the Bishnois of Rajasthan protecting *oraans* (contested by others for being 'too sparsely vegetated' to constitute a real grove); and the Mundas of Bihar mobilizing to prevent the destruction of *sarnas* by land developers by all available means, including the erection of a Christian cross outside a forest. One contributor working in the Western Ghats stressed the need to include a cultural-historical dimension resulting from the sanskritization of primitive gods, and the spread of the Aryan migration. Others were concerned by the contemporary significance of the groves by applying a 'stakeholder value importance index' to find out what current social functions and perceptions were attributed to the forests by villagers. Another added, however, that such values were not stable and could be modified with a linkage to the biosphere reserve concept, or be used to initiate new sacred groves in a Gandhian philosophy of self-sustainability, observing wryly 'after all, what is present today becomes history tomorrow'.

In one particular presentation entitled 'The logos and mythos of sacred groves', Saraswati went on to refer to the text of the Upanishads which he claimed had always considered the fig tree as a cosmic symbol rooted in the eternal god Brahman, asserting 'gods and trees are not separate from one another'. The names of sacred groves, he argued, 'do not mean sacred, which is a modern distinction that has come from anthropology', but refer rather to the individual names of deities, insisting that there was 'no such thing as the profane in traditional thought . . . the sacred-profane dichotomy is not applicable to any traditional society anywhere in the world'. Similar to Murtiraja, the professor felt empowered to generalize from the (alleged) situation in India to the status of any unitary 'traditional society'. Once again, to echo Douglas, the frame of reference substituted an anthropology world for a Hindu theology world.[11]

Drawing on the joint paper illustrating UNESCO case studies from Senegal, Sumatra and New Zealand, I presented the pre-prepared 'Synoptic overview of the diversity of the world's natural sacred sites'. Within the proposed comparative framework, I suggested, a distinction could be made between constructed monuments such as temples and churches; inorganic markers such as boulders, caves and mountain peaks made animate through symbolism and worldview; and organic features of a landscape considered sacred such as forests, groves and water springs. Memory-work within the landscape in its broadest sense, I also argued, could be quite idiosyncratic including many different layers of biographical, historical and mythological events (Hay-Edie 1999).

On the last day of the seminar, during a workshop to discuss potential case studies for the UNESCO comparative programme, Theomann later explained that, as an agnostic himself, he did not want to hurt anyone's feelings. Nonetheless, the main purpose of the seminar was 'not for the sake of preserving the sacred alone', but rather for conservation related to the ecological impact of pilgrimage, the legal status of sacred sites, and the rehabilitation of degraded lands. The enduring qualities of vernacular traditions appealed to UNESCO for their long-term ability to conserve biodiversity, legitimated by scientific criteria, on a foundation of customary values. Theomann feared, however, that there might be a worldwide trend in the erosion of faith, but on a more optimistic personal note added that:

> I see this in a cycle: faiths come and go. There is a great potential for faiths of coming back. In French we call it *fin-de-siècle* . . . people seek for some spiritual support and guidance. What I mean by that is that some spiritual values have a more lasting dimension, that go beyond purely financial or economic grounds.

As a yardstick, the utilitarian objectives of science and conservation therefore had to be seen as part of a wider context. Another senior Indian academic present, Mohan Ganesh, outlined an evolutionary perspective in which the history of the subcontinent progressed in four stages from pre-Aryan early hunter-gatherer existence; followed by a period of animism and small-sized sacred groves; through to hunting preserves coincidental with the advent of agriculture; culminating, finally, in an industrial stage marked by the formal creation of large national parks and international biosphere reserves. According to this historical trajectory, he also criticized the greater degree of reliance on codified regulations structured by 'rigid state bureaucracy'.

Reviewing the discussion, Ajay Ganga then asked if culturally recognized boundaries did indeed maintain 'effective institutions', could utilitarianism still be construed as the prime motivation for conservation? Characterizing the case made by Ganesh as one 'respective of collective norms' or 'cultural reverence', he juxtaposed this with the position of a social anthropologist, Mohammed Kalam (also present at the meeting as an observer) who had portrayed the continuity of sacred groves as resulting from the fear of retribution from a disturbed deity. Following Freeman (1994), Kalam had thus depicted the sacred sites agenda as a romanticist 'standard environmental narrative':

> Almost all the discourse on sacred groves revolves around preservation and conservation; the standard narrative being that because of devotion to gods/goddesses/deities there exist sanctions against entry, access to, and exploitation of resources in the sacred groves, and due to these sanctions certain groves have been preserved by people living in their vicinity . . . If people have (or to be more specific our ancestors had) innate faith in conservation where is the need for sanctions? . . . I feel the worship of a deity in an area of a forest evolved from the erstwhile practice of worshipping a patch of forest per se. Deities were installed at a much later date. The whole idea of worship was to propitiate a supernatural being or force which was thought of as controlling human destiny.
>
> (1996: 52)

Interestingly, while Kalam propounds a form of discourse analysis, he also slips into a logic of historical progress where innate faith based on fear is supplanted by a more recent religious dogma.[12] In Ganga's view, on the other hand, there was little need to invoke Kalam's 'innate faith in conservation', as sacredness should instead be considered to be at the very root of civic consciousness. More fundamental than stemming the loss of biodiversity, he argued passionately, the erosion of indigenous knowledge should be accorded top priority. Ganga, a university professor, thus encouraged the study of portfolios of sacred sites by 'building corridors of consciousness' and mobilizing diverse groups in society, including urban areas, around places of collective memory.[13]

In response, however, Ganesh then remarked that the sceptical political climate in Bombay during the 1970s had once treated him 'as a fellow who supports superstition' for his personal advocacy of 'flexible locally-adaptive strategies' for managing natural resources. Although centre-right governments had given their open support to the protection of sacred groves in the past, he cautioned that left-leaning politicians would consider such activities as superstitious, precipitating the following exchange with the UNESCO representative:

Ganesh: 'In Kerala, the debate between science and superstition is still very much alive.'

Theomann: 'Perhaps you can tell the states and authorities concerned with the issue of superstition that an international organisation, UNESCO, is backing this . . . '

Ganesh: 'Not in a communist state!'

Theomann: 'Well numerous communist countries are part of UNESCO: Libya, Cuba, North Korea . . . '

In terms of a conceptual network, the introduction of the word superstition introduced the critical issue of framing devices for any discussion of the sacred, thus revealing the common node around which translation across a variety of scales was taking place. The central concept of a sacred grove had been substituted for superstition, and then brought into contrast with communism. As political ideology, communism could then be generalized as variable across the Indian subcontinent, or perceived as homogenous within a single source of authority, UNESCO, an agency of the multilateral United Nations. The seminar thus provided a nexus of orders of comparison foregrounded together in the same event. Many of the participants encounter one another regularly at conferences, and some had been making presentations regarding sacred groves for years. The group of Indian scientists examining sacred sites had, in this respect, been momentarily nested within the global network soliciting case studies from around the world.

International animation

In his analysis of bureaucracy, Herzfeld reminds us of 'the capacity of local people to re-interpret official forms and invest them with meanings radically divergent from those of the law' (1992: 59). In much the same way, the concept of the sacred grove could not be definitively isolated at the meeting outlined above, neither as a result of the diversity and size of India; nor more particularly because set-aside areas of forest, as a product of community memory, fall outside any formal legal regime of protection. In the node of institutional connections present at the seminar, what constituted scientific or indigenous knowledge was made increasingly complex by a host of actors attempting to re-interpret versions of the sacred. UNESCO's symbolic involvement and the prospect of mobilizing other international donors' funds for conservation initiatives provided one such powerful framing device.[14]

During the international seminar, the interest in pursuing the UNESCO sacred sites proposal was taken up enthusiastically at the regional level (including proposals from Bangladesh, Sri Lanka, the Maldives, Iran and Mongolia), and efforts to put forward local case studies for funding were obvious during a field-trip to Iringole, one of the largest sacred groves in Kerala. Common to these diverse actors, however, was a symbolic mediation between scientific legitimacy and a perceived vernacular ability to conserve biological diversity, which as Theomann put it, arose 'under the common banner which is, of course, UNESCO'.

In terms of a globalizing set of symbols, UNESCO had thus adopted the role of global animator during the event and, in Theomann's own reasoning, had played a pivotal role in spreading a net over a wide catchment area to attract participants. The sacred groves of Kerala could indeed be said to have had a global significance like a cathedral in Rome. At the same time, numerous representations of sacred sites were also being exchanged during the meeting. Theomann concluded his institutional obligations by explaining that, before any follow-up work could take place, there would need to be further consultation with UNESCO National Commissions, MAB Committees and World Heritage state parties, while also stressing the 'importance of not demystifying any such site'.

Ironically, the seminar had been convened by UNESCO to collect case studies in the South Asian region, so what might constitute the de-mystification of a sacred site through the influence of an international institution had no obvious precedent. An assumption persisted that scientific knowledge, on the one hand, would be antithetical to traditional mystical practices and perceptions, on the other. As moderns, conservationists would unwittingly disenchant the primitives. Out of respect and cultural sensitivity for otherwise incommensurate systems of indigenous knowledge, UNESCO would therefore have to conceal its own activities, thereby denying its social efficacy. By stating that 'their reality' (local or indigenous people) was of critical importance for the protection of biodiversity, agency was thereby denied to the global body. The paradox is common, in fact, to many international organizations and NGOs promoting the concept of cultural diversity. A Mountain Forum report, for example, contains a section on the 'Sacred, spiritual and symbolic significance of mountains' that begins:

> Sacred values of mountains are of utmost sensitivity to people from these cultural and faith traditions. It should be absolutely clear that the study and understanding of religions or sacred sites is not universally an acceptable activity. Therefore, neither governments nor NGOs should presume to initiate or support any activities without first ascertaining that these are welcomed by local people and faith keepers.
>
> (1995: 23)

In this manner, a number of international conservation organizations have recently become intrigued by the concept of sacred sites and are endeavouring to revitalize locally situated approaches to sustainable development. Yet, the attribution of agency and legitimacy in the network of relations is fraught. The Mountain Forum states tautologically that it is 'absolutely clear' that a project designed by one set of actors is 'not universally an acceptable activity': a universal intellectual rule, purportedly, that universality cannot ever be implemented. Similarly, UNESCO's objective to work within certain utilitarian parameters (that of linking sacred sites with biodiversity conservation) also set out explicitly to avoid any imposition of its own scientific models on local people. Indeed, the global organization had to partially mask the fact that it perceived itself as initiating such activities, despite the manifest lobbying by indigenous peoples at the United Nations WGIP for the recognition of their ancestral lands as sacred.

Ultimately, the process of animation in an international network can cut both ways. A United Nations agency re-affirms the 'vernacular conservation' of biodiversity, while participants presenting case studies of indigenous knowledge at a regional seminar respond as willing co-animators by colouring in the empty global category. The contribution of UNESCO may thus unexpectedly amplify the symbolic scope for diverse sacred sites. The collision of two or more formerly unconnected structures around a common formula of association, such as the notion of the sacred grove, may in this way animate a disparate set of actors concerned with the sustainable use of biological diversity.

Notes

1 For reasons of access as well as anthropological interest, issues pertaining to the use of funds, budgets and institutional management within UNESCO (allocated at the biennial General Conference) are not discussed here. UNESCO has, however, had a chequered political history marked by the withdrawal of the USA, Singapore and the UK in the mid-1980s for ideological and administrative reasons.

2 WWF organized the first Summit of Religions in Assisi in 1986 resulting in Faith and Nature declarations by Buddhists, Christians, Hindus, Muslims and Jews. This was

followed by a second summit in Windsor in 1995 drawing together a total of nine faiths with the addition of the Bahais, Sikhs, Jains and Taoists. Each faith produced a book as part of the WWF Ecology and Faith series (Batchelor and Brown 1992; Pedersen 1995).

3 A number of documentary references drawn from botany, ecology and anthropology were circulated for UNESCO staff to read. For example, in an analysis of Ficus species in Vietnam, Dinh (1996–1997) commented that in ethnoscience '*L'arbre-repère est aussi l'arbre-repaire*' ('reference trees are also repair trees'). Similarly, a film proposal was circulated concerning fig trees which in many parts of the world have symbolic potency either as images of fertility, fecundity or dangerous natural exuberance.

4 During earlier preparatory meetings the anthropologist had voiced a concern that (in her eyes) the UNESCO logo associated with the conference would 'not give a good impression to the academic community' and wanted to restrict the participants to a 'rigorous group of professional researchers'. She insisted, furthermore, that the term 'cultural integrity' be dropped from the title as it would be meaningless to social anthropologists.

5 On the first day of the WGIP in 1997, the indigenous delegates re-enacted the visit of the first native American chief Deskaheh to the then League of Nations in the 1930s. As a procession with a buffalo skull held by five of the original indigenous representatives at the working group, about 400 delegates from all over the 'fourth world' in ceremonial dress converged on the plenary hall, followed by camera crews (including myself with a video camera), where prayers and songs (including an eagle/bear rendition by a Siberian shamaness) embodied their political claim to 'indigenousness'.

6 Article 8j of the CBD affirms the obligation for national governments to: 'Respect, preserve and maintain knowledge, innovations and practices of indigenous and local communities embodying traditional lifestyles relevant for the conservation and sustainable use of biological diversity'.

7 Material from this seminar was recorded on video, notes taken by hand as well as in some cases verifiable in the form of transcripts of speeches. The official proceedings of papers with diagrams and references appeared as a book published by Oxford-India entitled Conserving the Sacred for Biodiversity Management (1998).

8 IHDP is a joint programme of the International Council of Scientific Unions (ICSU) and the International Social Science Committee (ISSC) with a central office in Bonn, Germany, aiming to provide a 'bridge between the biophysical aspects of global change and human dimensions' on issues such as biodiversity and climate change.

9 The concept of the biosphere was first coined by the geographer Vernadsky as the thin layer of all life on the surface of the inorganic planetary mass, and gained prominence with the activities of the international Man and Biosphere (MAB) programme of UNESCO launched in 1973.

10 Douglas makes use of a reference to Nelson Goodman's (1978) Ways of Worldmaking who first argued 'that the rightness of categories depends on their fitting within a world' (1987: 17).

11 Saraswati nonetheless relied on an anthropological reference concerning the lack of a sacred-profane distinction amongst certain Australian Aborigines. The prognostic has echoes of Obeyesekere's critique of Marshall Sahlins' interpretation of Captain Cook as a god, which he deemed eurocentric – as Sahlins observed of Obeyesekere, Saraswati seemed to attribute his own South Asian identity with a more legitimate claim to a 'traditional point of view'.

12 Mitra and Pal adopt a similar argument in a review of the status of sacred groves across the subcontinent, invoking James Frazer's timeless Golden Bough which they

observe 'written in 1935, narrated how people, right from Palaeolithic times, preserved forests by worshipping them. In these forests, no tree could be axed, no branch broken . . . ' (1994: 22). The authors are quick to show, however, that in contrast to a mythical past, the majority of the remaining sacred groves, as 'vestiges of an ancient practice', are under intense development pressure through the loss of traditional cultural values.

13 In the form of an existing knowledge network, disseminated in magazine format, social entrepreneurs such as healers and herbalists, different faith perspectives and small-scale technological innovators, should thus be able to contribute incrementally towards the promotion of the livelihoods of local communities.

14 As Chandrakanth and Nagaraja (1992) comment elsewhere in a discussion of the social fencing of sacred groves in the Indian state of Karnataka, 'the village community . . . the government, political organizations and the judiciary have great responsibility in preserving the *devara kadu* institution in Coorg which is unique in the world' (1992: 222–223). What seems clear with the mention, once again, of the world as a frame of reference, is a high degree of international attention to the potential benefits of animism and indigenous knowledge for biodiversity conservation.

References

Batchelor, M. and K. Brown (eds). 1992. *Buddhism and Ecology*. WWF World religions and ecology series. London: Cassell.

Chandrakanth, M.G. and M. Nagaraja. 1992. Existence value of Kodagu sacred groves: Implications for policy. In *The Challenge of the Ecological Balance*. New Delhi: Centre for Science and the Environment.

Dinh, T.H. 1996–1997. Signes-nature, signatures, biodiversité (1): les bosquets cultuels au Vietnam, pour un concept de vestiges verts. *Cahiers d'Etudes Vietnamiennes*. 12. Paris.

Douglas, M. 1987. *How Institutions Think*. London: Routledge and Kegan Paul.

Ellen, R.F. 1986. What Black Elk left unsaid: on the illusory images of Green primitivism. *Current Anthropology* 6 (2): 9–13.

Escobar, A. 1994. *Encountering Development: The making and un-making of the Third World, 1945–1992*. Princeton: University Press.

Freeman, R. 1994. Forests and the folk: perceptions of nature in the swidden regimes of highland Malabar. *Pondy Papers in Social Sciences*, 15. India: Institut Français de Pondicherry.

Gell, A. 1999. *The Art of Anthropology: Essays and diagrams*. LSE monographs on social anthropology. London: Athlone Press.

Hay-Edie, T. 1999. Landscape perception and sensory emplacement. In *Cultural and Spiritual Values of Biodiversity: A complementary contribution to the global biodiversity assessment*. (comp. and ed.) D.A. Posey. United Nations Environment Programme. London: Intermediate Technology publications.

Herzfeld, M. 1992. *The Social Production of Indifference: Exploring the symbolic roots of western bureaucracy*. New York: Berg.

Kalam, M. 1996. Sacred groves in Kodagu district of Karnataka (South India): A socio-historical study. *Pondy Papers in Social Sciences*. 21. India: Institut Français de Pondicherry.

Little, P. 1995. Ritual, power and ethnography at the Rio Earth Summit. *Critique of Anthropology* 15 (3): 265–288.

Mitra, A. and S. Pal. 1994. The spirit of sanctuary. *Down to Earth* (31 Jan.), 21–36.

Mountain Forum. 1995. *Report of the Initial Organizing Committee of the Mountain Forum.* Convened by the Mountain Institute, West Virginia. Unpublished.

Mutoro, H.W. 1994. The Mijikenda kaya as a sacred site. In *Sacred Sites, Sacred Places.* (eds) D.L. Carmichael *et al.* London: Routledge.

Parkin, D. 1991. *Sacred void: Spatial images of work and ritual among the Giriama of Kenya.* Cambridge: Cambridge University Press.

Pedersen, P. 1995. Nature, religion and cultural identity: The religious environmentalist paradigm. In *Asian Perceptions of Nature: A critical approach.* (eds) O. Bruun and A. Kalland. Richmond: Curzon Press.

Perez de Cuéllar, J. (ed.). 1995. *Our Creative Diversity: Report of the World Commission on Culture and Development.* Paris: UNESCO Publishing.

Ramakrishnan, P.S. 1996. Conserving the sacred: From species to landscapes. *Nature and Resources* 32 (1): 11–19.

Roussel, B. 1992. A propos de l'historicité des forets sacrées de l'ancienne côte des esclaves. In *Plantes, paysages et histoire en Afrique Sud Saharienne.* Paris: Karthalo-ORSTOM-CRA.

Schaaf, T. 1995. Sacred groves: Environmental conservation based on traditional beliefs. In *Culture and Agriculture: World Decade for Cultural Development 1988–1997* orientation texts. Paris: UNESCO publishing.

Siebert, U. 1996. The meaning of the category 'indigenous' within the United Nations: The politics of the debate on a definition. Masters dissertation. University of Berlin, Unpublished.

UNESCO. 1995. The cultural dimension of development: Towards a practical approach. *Culture and Development* series. Paris: UNESCO Publishing.

Wright, S. 1998. Encaging the Wind: conference review of 'The Power of Culture – UNESCO international conference on cultural policies for development', Stockholm (March–April 1998). *Cultural Policy* 5 (1): 173–182.

Chapter 8

The globalization of indigenous rights in Tanzanian pastoralist NGOs

Greg Cameron[1]

In this chapter I focus on a Tanzanian pastoralist network called the Pastoralist Indigenous Non-Governmental Organizations (PINGOs) Forum where the original vision of the affiliate NGO members was undermined in part by the top-down programmes and structures of Western donors who were imposing notions of indigenous human rights from other international contexts. Section 1 examines the capture of the PINGOs Forum Secretariat by the leadership of one of the founder affiliate members, a process which saw the PINGOs Forum veer away from its original mandate of membership networking, advocacy and lobbying the Tanzanian government, to the international arena. Section 2 analyses the origins of indigenous rights processes and their potential decontextualization within the PINGOs Forum. Section 3 examines the murkier side of indigenous human rights initiatives within the PINGOs Forum, delineating the ways in which the PINGOs leadership, in practice, employed notions of the 'indigenous' in an exclusionary way so as to buttress their power base. Section 4 situates indigenous human rights, and donor fundraising and advocacy in the West, arguing that such perspectives exoticized local communities while having virtually no positive impact on the mission of PINGOs or the affiliate members. Section 5 argues that the communitarian model of society-state relations as propounded by some donors, and supported by the PINGOs leadership, was overly simplistic and based on notions of 'dominant' and 'dominated' groups and a monolithic state.

The emergence of the PINGOs Forum

In Tanzania, pastoralist and hunter–gatherer lands had been feeling the most immediate pressures since economic liberalization in the 1980s and 1990s, due to schemes that largely ignored their traditional land rights, whether it was from state farms, conservation interests, private agribusiness or in-migration by small scale agriculturalists.[2] Among pastoralists' organizational responses to land alienation, my focus is on the emergence of pastoralist community-based organizations (CBOs) and non-governmental organizations (NGOs). The origins of these NGOs, mainly Maasai, varied, and encompassed a number of diverse forms including one-man shows, where NGO initiatives and registration were ideas hatched by individuals; group ranching concerns, where land acquired was used to engage in commercial enterprises; peasant-pastoralist memberships undertaking rural development projects, along with conflict resolution activities; and educational activities centred on vocational training. In some cases they were originally CBOs in existence for some years. For example, in the case of the Barabaig, the impetus was the extreme conflict in Hanang district around the wheat farms of a Tanzanian parastatal (NAFCO) from the 1970s, which were financed by the Canadian International Development Agency (CIDA).[3] There were also representatives from Tanzania's hunter–gatherer community, the Hadzabe.

Mobilized and registered during the beginning of the transition to political and economic liberalization in the early 1990s, these pastoralist organizations worked hard together advocating nationally and internationally on the problems of community land alienation. By 1994, these disparate pastoralist NGOs felt confident enough to establish the PINGOs Forum. In 1995 the PINGOs Forum decided to establish a centre in Arusha Town in order to coordinate the widely scattered members throughout Tanzania's northern dry lands. The centre contained two office rooms, a meeting room, a small library and resource centre, and sleeping and dining facilities for visiting members. A secretariat, joint committee, and general assembly, together with sub-committees, were activated. The members of these bodies were elected or hired, minus some secretariat staff, from the respective CBO/NGO affiliate members.

Stating its overall objective as the strengthening of Tanzanian pastoralists and indigenous communities including the Maasai, Barabaig, Hadzabe and Il-Dorobo peoples,[4] the PINGOs Forum envisioned defending marginalized peoples in Tanzanian society by supporting its NGO affiliate members in their efforts to strengthen and protect indigenous culture and knowledge, promote socioeconomic development, protect the environmental habitat of indigenous peoples, and defend the human rights of their member constituencies as citizens with full rights. PINGOs also aimed to promote relationships with other national and international organizations, or groups of persons advocating the welfare of Tanzanian indigenous minorities.[5] In fact, PINGOs saw itself as the overarching organization of the fledgling pastoralist movement in Tanzania. Demographic and geographic horizons would widen further still as PINGOs came to define itself as the defender of East Africa's 'indigenous peoples' – some of its personnel would even set up a duplicate Africa-wide indigenous apex, which I detail below – thus subsuming the category of 'pastoralist' within its 'indigenous' representational ambit.

The early programmes and directions chartered by the secretariat enjoyed the confidence of both donors and affiliate members, numbering approximately nine CBOs/NGOs throughout the period under study, 1996 to 1998 (though several other groups, mainly Maasai, joined towards the end of this period).

PINGOs demonstrated the benefits of self-organization, doing good work in establishing a radio call for meetings and emergencies, or advocating to the media on the economic benefits of pastoralism to the national economy. PINGOs also made tentative links with a national land rights NGO in Dar es Salaam, which was organizing for the defence of community land rights.[6] Meetings at both the level of the board and general assembly were also briefly regular. This cohesive programming and unity was not to last long due to dynamics within the affiliate membership. Aside from the tension between mobilization and service provision in the communities of the affiliate NGOs, the leaders of numerous founder affiliates, after receiving donor funding, failed to keep their memberships informed of their activities, causing suspicion of mismanagement of funds. The ensuing conflict saw many leaders removed and members resigning, thus imploding and dividing their community organizations. These problems affected key PINGOs affiliates and made them unable to fully participate in the activities of the PINGOs Forum in Arusha. The more independent and powerful founder NGOs became internally divided and thus institutionally ineffective which, coupled with weaker new members devoid of funding, left a power vacuum that was quickly filled by another founder NGO, Illaramatak and its offshoots (meaning that Illaramatak leaders sat on the boards of these other NGOs). The coordinator and the chairman of PINGOs both hailed from Illaramatak. Due to his earlier activism on behalf of his community in his home district of Simanjiro, the coordinator had attracted numerous donors who began rushing in with quite different types of programmes. PINGOs rapidly centralized its bureaucratic and financial systems. Thus from the original vision of the founder organizations, which was for PINGOs to be a simple coordination centre meant to link members, lobby the government and network with international bodies, PINGOs rapidly evolved into a multisectoral NGO competing with its affiliate members for donor funding for various activities: conducting community paralegal training, implementing rural development projects, facilitating court cases, participating in workshops and attending international conferences. There was a drastic suspension of internal meetings between the secretariat and board and general assembly. Non-transparent financial systems and poor staff morale also afflicted its institutional cohesion and further exacerbated the secretariat's maladministration. The international orientation of the PINGOs Forum towards advocacy in the West was particularly deleterious to the implementation of its original mission. Given the intractability of the problems with the Tanzanian government, it was understandable that the PINGOs leadership chose the line of least resistance and channelled their energies into the more accommodating international arena. But this external orientation was as much to do with the difficulties of engaging with the Tanzanian state as it was with the unaccountable nature of the PINGOs leadership, a faction that eagerly linked up with equally eager donors seeking North–South partnerships with this in-vogue pastoralist-indigenous apex.[7]

This chapter is about the political economy of knowledge surrounding the relationship between local and international responses to pastoralist land problems. It highlights concerns not unlike those of J. Pottier's work on eastern Democratic Republic of Congo, whose inhabitants were being denied expression when dominant interests demanded that local views be replaced by more simplistic perspectives.[8] Although this chapter details dynamics within an NGO sector, rather than the state level during time of war, there are parallels in the ways in which certain forces claimed to have a knowledge superior to that articulated by local voices, a knowledge backed by power and giving full play to the murkier side of everyday forms of power struggles over what constituted indigenous knowledge. PINGOs Forum is an important case study in the rise and fall of pastoralist NGOs. Certainly it was the only network of the pastoralist NGO sector in Tanzania, and one of the few pastoralist apex organizations on the African continent. The potential for PINGOs to play a minor yet important role in facilitating community responses to pastoralist and hunter–gatherer concerns seemed boundless in the beginning. That PINGOs fell abysmally short of its laudable goals, amidst recrimination, betrayal and opportunism, was in large measure due to donors' framing of the structures and processes for pastoralist community responses to the problems of land conflict and socioeconomic development. Marginalized, were the plurality of voices amongst the affiliate members who clearly lost control of the trajectory of an organization that they themselves had founded based on their own community priorities. The impact of the global upon the trajectory of the PINGOs Forum was to increasingly swerve the organization away from its mission to defend the land and cultural rights of Tanzanian pastoralists and hunter–gatherers and instead to focus its demands for economic and cultural entitlement as indigenous Tanzanians. My concern is less with trying to define indigenous in the African context, or even questioning the analytical validity of the term indigenous in Africa, sceptical as I am. Neither am I addressing the very real advances made by indigenous communities in other contexts of the globe, both locally, nationally,[9] and internationally at the United Nations level. Nor, in fact, am I critiquing the work of the NGOs mentioned here in their larger programming contexts, unless specifically highlighted. Rather, my concern is with how international indigenous rights, as an aspect of indigenous knowledge, was adapted and actualized inside of the PINGOs Forum, and the implications of this international framework for the forms and possibilities of self-organization in impoverished communities resisting top down developmentalism in Tanzania.

The (de)contextualization of indigenous rights in the PINGOs Forum

Affinity with indigenous peoples worldwide was evident in the PINGOs Forum ranging from conferences and funding proposals to development programmes and exchanges. One PINGOs Forum conference schedule noted 9 August, the

UN International Day for Indigenous Peoples, as an annual celebratory event held collectively throughout PINGOs, as well as individually in every affiliate member's area. PINGOs documents and proposals to donors claimed it would be a time to express solidarity with other indigenous peoples everywhere.[10] Literature deposited in the PINGOs Documentation Centre spoke of the workings of the Saami parliament in the Nordic countries, as well as successes in Canada around Inuit land claims. But what was meant by the term indigenous? Who was employing the term and why? Where did the term come from and in what context was it originally employed? In attempting to address these questions this section examines briefly the origins of international indigenous human rights. Next it turns to the often tortured efforts of intellectuals and NGOs to apply the concept to the complexities of a rapidly changing African continent.

Common in the literature on indigenous peoples is that the territorial claims of indigenous peoples may be rooted far back in history and are underpinned by the notion of their special claims to the land; their unique relationship with the environment is crucial to their survival, and their land and resources may never be ceded.[11] Observers concur that the movement for indigenous rights began in the West, among American and Canadian native peoples, and including Inuit, Saami, Maori and Aboriginal peoples from Europe and Australasia. Land rights began to receive attention in the United Nations (UN) and its agencies in the 1950s. The first international legal instrument to codify indigenous and tribal people's rights was the International Labour Organization's (ILO) convention concerning the Protection and Integration of Indigenous and other Tribal and semi-Tribal Populations in Independent Countries (1957). In 1989, the ILO adopted a new Indigenous and Tribal Peoples Convention that emphasized people's rights to control their own development.[12] Land rights were central to the Draft Declaration on Indigenous Rights issued by the UN Working Group on Indigenous Populations (WGIP), established in 1982. At present, there is a process in Geneva, including the UN Draft Declaration on the Rights of Indigenous Peoples which affirms the right of indigenous peoples to determine their own development; and the obligation of states to obtain informed and free consent to any project affecting indigenous people's territories.[13] Indeed the whole concept of indigenous peoples is now inseparable from the human rights discourses that represent them as victims of abusive governments. As Wilson points out, indigenous peoples in contexts as different as Panama, Canada and South Africa have engaged in close negotiations with their own governments over their constitutional claims for linguistic and territorial rights and political sovereignty; and more soberly adding that it is no coincidence that activities around the UN Year of Indigenous Peoples in 1993 came under the UN's human rights budget line, though the amount was meagre.[14]

The WGIP was open to all, especially as self-definition was considered paramount in being able to partake in the process, with no one having the right to check the authenticity of another group. However, matters came to a head in 1995 when Boers, defining themselves as 'indigenous' (from what many considered to be the former dominant group within the South African apartheid system) entered the WGIP process. The controversial, albeit temporary, entry of the Boers into the WGIP catalyzed the beginning of a process to more rigorously define indigenous in Africa.[15] Within WGIP there was a limit to how inclusive the movement could be without loosing its sense of direction, a feeling further triggered by the growing participation of African representatives in the WGIP. In 1998, A. Martinez, a UN special rapporteur, along with Native American representatives, argued that it would be more appropriate that African and Asian delegates submit their grievances to the UN Working Group on Minorities. This was rejected by the majority in the WGIP, but nonetheless galvanized African delegates to begin defining indigenous in the African context.[16]

Numerous observers have sought to conceptually clarify the implications of the UN process around international indigenous rights for Africa. Thornbury, for one, argues for more clearly defining indigenous at the UN level, citing advantages such as improving the goodwill of governments; giving confidence to indigenous people; and improving precision in targeting programmes. Correctly pointing out, though not addressing in detail, the obvious potential disadvantage if there were to be positive outcomes for indigenous peoples internationally, Thornbury acknowledges that there could arise 'sundry collectivities' of peoples with different kinds of objectives reclassifying themselves accordingly, a practice which he terms 'indigenism'.[17] Nonetheless sympathetic observers on the indigenous rights issue have argued that sooner or later the issue of definition has to be addressed in Africa; while others have gone straight to the point, as it were, seeking a common definition that, if nothing else, can serve as a temporary conceptual handle while practice on the ground further fleshes out the meaning of indigenous in Africa. This work-in-progress definition stresses that indigenous identity is a social reality experienced by peoples distinguished in myriad ways, and who are perceived to be different from the dominant society; indeed a concept is needed in international law to describe such sections of a population and their position as indigenous peoples in relation to (politically and numerically) dominant sections.[18] Indigenous peoples are thus conceived as being internally colonized by their post-colonial nation-state and society's dominant groups. The core feature of this relationship is the lack of recognition by the nation state of the distinct background and special needs of indigenous peoples.[19] Saugestad, for instance, sees the concept as both sociological and legal: where activists can look at declarations made in international fora and take them back as levers for reforms in their respective national contexts.[20] In Africa, moreover, a more specific definition vis-à-vis the local has been made by a continental-wide indigenous apex which defines indigenous peoples as those who maintain a distinct culture, a historical continuity with land, and who are discriminated against, marginalized or displaced in their own countries.[21] Yet at the same time, this very same apex acknowledges that no accurate figures exist yet on the number and size of communities claiming indigenous status in Africa, estimating that 27 African countries have indigenous communities (i.e. those claiming status within the UN framework).[22] Turning to East Africa, Maasai leaders, among others, were increasingly becoming aware of international fora to advocate their position, in contrast to the closed corridors of power at the national level in Tanzania and Kenya.

How does one translate an international concept to fit emic feelings of dispossession unless an interlocutor actively seeks to introduce such a concept? Can an international concept have validity in a local context? Were new identities being actively promoted based on transnational priorities? International NGOs in some cases reflected sincere, if misguided attempts, to help certain Tanzanian communities under incredible socio-economic stress, especially around land dispossession. In other cases, however, there were political agendas at play, which we examine below. In either case, donor interventions had the potential to socialize local NGO pastoralist leaders in a new internationalized discursive framework, a process that would take place both at home and abroad, and which would have implications as to how, where, and with what identity, they would represent their local constituencies. Before examining this relationship between the PINGOs Forum and its donor partners, the following section will address the increasingly exclusive nature of indigeneity inside of the PINGOs Forum itself.

The exclusivity of indigenous rights inside of the PINGOs Forum

The severe clashes with the Tanzanian state, such as the case of the Barabaig in Hanang, as well as clashes between Maasai pastoralists and small and large agriculturalists, along with the institutional presence of donors within the PINGOs Forum, was catalytic in creating an awareness of international indigenous rights law inside of PINGOs.[23] Why did the PINGOs leadership so willingly employ notions of indigeneity to themselves, and dominant groups to other Tanzanian communities? How were indigenous human rights absorbed into the PINGOs programmes and with what consequences? How indigenous were PINGOs' ethnic groups in the affiliates?

In Arusha, in the PINGOs Forum Secretariat, the leadership never defined indigenous, perhaps given that the word was part of its institutional name. This lack of clarity is not surprising given that even international bodies like the UN or the European Union's policy unit on indigenous peoples had yet to satisfactorily define indigenous in the African context. The UN Cobo definition emphases pre-invasion/pre-colonial societies found in the Americas and Australasia, where there was a clear-cut white colonial annexation.[24] In northern Tanzania, the Maasai and Barabaig are recent historical newcomers compared to neighbouring Bantu, Cushitic and Hadzabe peoples inhabiting northern Tanzania. Others have questioned the indigenous credentials of the Maasai, less so because of their being latecomers to their present habitat, but more because they have reached a level of organizational accomplishment and diversification within their groups that has brought considerable wealth and political influence in some sections. The political representation achieved in government, including among members of parliament and in cabinet in both Kenya and Tanzania, would also bring into question the extent of their marginality from mainstream East African society.[25] Yet PINGOs purported, or aspired, to represent great swaths of East African peoples. An early contact with the International Working Group on Indigenous Affairs (IWGIA) saw PINGOs asking to be the first IWGIA national group in Africa, and submitting a two year budget of US$84,000 for allowances, office equipment, capital expenses, a four-wheel-drive vehicle, and a plot of land.[26] The rationale of this proposal to IWGIA was centred on the defence of the rights of indigenous organizations, not only of pastoralists, but hunter–gatherers, throughout East Africa who were advocating self-determination for indigenous peoples. An IWGIA office in Arusha would serve, so the correspondence went, as an educational function in raising the public's awareness of the issues facing indigenous peoples in Africa. It could also raise the profile of the Maasai, Barabaig, Hadzabe and Il-Dorobo peoples in Tanzania; the Turkana, Samburu and Maasai peoples of Kenya; the Karamajong of Uganda, as well as other marginalized ethnic minorities.[27] Some of the PINGOs staff had in fact never heard of some of these peoples, including the Il-Dorobo of Tanzania.[28] Yet the PINGOs leadership presented itself as a beacon of indigenous rights in East Africa, and more than willing to ally itself with transnational organizations in actualizing its vision of indigeneity. The request was not funded for the time being, though IWGIA would offer its own terms of programme partnership with PINGOs, an issue which is addressed below.

What did affiliate members think of indigenous rights? The extent to which the concept was internalized by affiliate members, or adapted in certain ways to their predicament in their communities (a kind of last ditch effort at mobilizing outside allies) is not evident or whether it was merely addressed to the donor community for instrumental purposes by affiliate leaders. Undoubtedly there was an early awareness of international indigenous rights and international legal principles at all levels of PINGOs. Various PINGOs documents for the affiliates cited the UN Declaration of the Rights of the Indigenous Peoples; the ILO Convention 169 on Indigenous and Tribal People; and the African Charter. Laws were cited as guaranteeing the cultural right to freedom of worship and traditional spirituality for ethnic and religious minorities.[29] One PINGOs proposal requested funding for the translation of ILO Convention 169, and the UN Declaration on the Rights of Indigenous Peoples, into the Maa, Barabaig and Hadzabe languages (the documents had earlier been translated into Swahili).[30] An early letter from a Barabaig member NGO to a UK donor requested US$7,000 for an international conference on human rights, culture and development, the justification being that, given that 1993 was the International Year for the World's Indigenous Peoples, it would be ideal to promote the plight of the Barabaig.[31] Another affiliate had under its budget line for special events: the International Day for Indigenous People (August 9);

Human Rights Day (December 19); International Women's Day (March 8); the anniversary of their NGO; and finally, Water Festival Day. The preamble of this affiliate NGO's constitution, incorporating the Tanzanian constitution, reaffirmed the right to culture, land, natural resources, as well as international legal instruments, including the UN Declaration of Rights for Indigenous Peoples.[32]

Turning to the actual relationships in PINGOs, tensions became manifest in the affiliate membership over the state of the leadership in the secretariat, first among non-Maasai affiliates, generally speaking. Bulgalda, the Barabaig member, dropped out of PINGOs due to internal leadership problems, and also perhaps due in part to the suspicions that the Barabaig NGO members had of the Maasai leadership inside of PINGOs. The Bulgalda leader pointed out that the Maasai and Barabaig had a long tradition of enmity dating from the nineteenth century. There was also the fear of a hidden agenda of the PINGOs leaders and that the confrontational approach of the Maasai leaders would ultimately be at their expense.[33] Even the leader of the remaining Barabaig member NGO made it clear on numerous occasions that the Maasai leadership under the coordinator did not represent him or his community. In one instance, the secretariat produced the first newsletter on behalf of the PINGOs Forum that was titled in Maasai. The members at this particular meeting, both Maasai and non-Maasai, vetoed the newsletter title, though suspicions among the Barabaig delegates may have persisted.[34]

In the case of the Hadzabe there was no NGO, and indeed, Hadzabe participation in the dormant governance structures of the PINGOs Forum was from the beginning more apparent than real. Meanwhile Hadzabe communities were facing a crisis as Barabaig, forced from their land in Hanang district due to the CIDA-financed NAFCO wheat farms cited above, began competing for water resources for their cattle against the wildlife the Hadzabe depended upon for hunting. In-migration from neighbouring Iraqw agriculturalists and nearby commercial hunting blocks also contributed to the depletion of Hadzabe game. It is believed that the Hadzabe number approximately 1,000 people spread over the districts of Singida, Karatu and Mbulu. Hadzabe hunter–gatherers are the poorest of the poor in terms of exclusion from wider economic and political processes as well as recognition by wider Tanzanian society – including pastoralists – of their livelihoods as a genuine way of life that need not be transformed into another mode of economy such as agricultural cultivation. The Hadzabe, without their own organization, were by default to have PINGOs as their primary NGO, unlike other members who were institutional affiliates with their own NGOs/CBOs. This, despite there being literate Hadzabe youth who could have formed their own CBOs to represent their hunter–gatherer community.[35]

What makes some knowledge, people and settings more indigenous than others? How is indigeneity (mis)used in particular contexts? The situation of the Hadzabe and Barabaig within PINGOs, and their marginalization from the decision-making processes within the secretariat, highlighted the exclusive and multiple meanings of indigenous even within PINGOs. It is not clear how seriously the PINGOs leadership thought of themselves as indigenous – either in an international or local sense – relative to other Tanzanians. It would appear that the leadership was making strategic decisions in the light of power relations and access to resources, for not only were non-Maasai shut out of the decision-making processes but even most Maasai were excluded, particularly those who dared challenge the PINGOs leadership. Even the justification of Illaramatak's dominance of PINGOs over the other affiliates was couched to the main funder by the PINGOs leadership as analogous to the US dominance of the United Nations' system. While the analogy could refer to the perception by the PINGOs leaders of the benign moral and financial support that the US supposedly lent to the UN, an alternative interpretation could decode this statement as one justifying hegemonic dominance over an organization that served as a tool for the dominant member, Illaramatak, a view held by most of the PINGOs affiliate membership.[36] The Illaramatak leadership controlling PINGOs viewed their domination as benign and positive, and possibly saw validity in the analogy from its exposure to international indigenous rights processes. The majority of members, on the other hand, came to see PINGOs as merely another offshoot of Illaramatak, or as another example of a charismatic pastoralist NGO leader unable to make the transition to becoming an NGO manager.[37] By 1997, the secretariat had become a narrow Maasai grouping purporting to speak for a wider indigenous constituency, employing indigenous rights dialogue selectively, and backed by international advocacy NGOs. Halfway through the UN International Decade of Indigenous Peoples, there were problems at PINGOs as the main Dutch donor, Novib, initiated an external evaluation of the PINGOs Forum.

Donor NGOs and the PINGOs Forum

Throughout 1997 to 1999 PINGOs received a constant infusion of finances from donors. The financing of these pastoralist NGOs/CBOs and PINGOs, including the rent, meetings, furnishings, computer equipment and the salaries for the secretariat in the Arusha offices, included a wide assortment of International NGOs from Western countries, particularly the Netherlands, Canada, Britain and the Nordic countries. Recurrent expenditures, in particular, were met by the Dutch NGO, Novib.[38]

Western NGOs, generally speaking, considered PINGOs and its affiliate members indigenous. For some donors, the indigenous question in Africa resonated with the indigenous politics of their native countries, including the struggles of the First Peoples in the Americas and Australasia;[39] and the Greenland Inuit and the Scandinavian Saami for the Nordic donors. Inside the PINGOs Forum Secretariat itself, the relevancy of the international indigenous concept bobbed and weaved according to the vagaries of the internal dynamics of the organization, eventually assuming importance as partnerships with older international NGOs began to come under strain. Consequently other organizations more geared to the needs of tribal peoples internationally became paramount and made a more determined effort to actualize the international concept of indigenous within the forum and through it, into the wider pastoralist world. In short, indigenous politics did not immediately manifest itself in PINGOs; only later did notions of indigeneity assume greater importance as PINGOs ossified, and as traditional donor partners either began to reconstitute themselves, like the increasingly cash-strapped Canadian University Services Overseas (CUSO), or sever the partnership with PINGOs, in the case of the well-financed Novib.

I was employed by CUSO and will address some of the dynamics of concern in this chapter from inside of CUSO.[40] CUSO has been in Tanzania since the mid-1960s with programmes in a number of countries in Asia, Latin America and Africa. Its global themes of sustainable economic alternatives and the cultural survival of indigenous peoples are actualized through the placing of volunteers, termed cooperants, who seek to facilitate links between social movements in the South and North. CUSO initially provided basic infrastructural support to the PINGOs Forum. CUSO had also played a role in assisting a number of the affiliate members of PINGOs before pastoralist NGOs had attracted a wide number of donors, including the Barabaig, the Hadzabe and some Maasai CBOs. In regard to the Barabaig, a kind of family feud within the development community had broken out back in Canada. CUSO had called the Canadian International Development Agency (CIDA), the bilateral agency of the Canadian government and CUSO's main financer to boot, to task for sponsoring the Tanzanian food parastatal's expulsion of the Barabaig in order to set up wheat farms in Hanang District.

CUSO also facilitated the linking of Barabaig and Maasai activists, from the future affiliate NGOs of the PINGOs Forum, into educational and activist quarters in Canadian Aboriginal organizations. One example was at a workshop on the nature of the economy of Canadian Aboriginal communities in the province of Ontario, where there were in attendance a Maasai and Barabaig representative.[41] Certainly the exchange of ideas and exposure would have further heightened their direct experiences. Delegates were in fact hopeful that the gathering had created a momentum that would carry through to the United Nations. The Barabaig representative told the delegates that he would return to Tanzania encouraged to continue the struggle against land loss and World Bank policies that were re-colonizing their countries and undermining their indigenous culture.[42] Canadian Aboriginal delegates expressed their belief that international linkages had been the exclusive domain of corporate, government and NGO leaders for some time: what could be more promising than the beginnings of dialogue among the world's first peoples? Delegates also visited a fish farm and a business development corporation, then issuing five million Canadian dollars in loans. Nonetheless, despite the real achievements of Canadian Aboriginal communities, and the exchanges of this workshop, there would have been few concrete parallels for the Tanzanian delegates to take home between a powerful capitalist economy and an economically underdeveloped authoritarian state. Nor did the NGOs, from which the Barabaig and Maasai delegates originate, have the years of experience and institutional capacity to filter and usefully adapt such an overseas trip to the advantage of their community organizations. And increasingly, as CIDA slowly cut the global CUSO budget from the early 1990s, CUSO itself lacked the capacity to work with such activists to process these overseas experiences and fit them somehow, both methodologically, conceptually and politically to their local community organizations in northeastern Tanzania.

Before long in fact, the CIDA budget cuts catalyzed a programmatic shift within CUSO to less political and more opaque themes such as human rights and gender equity where any number of placements and types of local NGO partners could be encompassed within the organization's ambit, including indigenous human rights. This trend was moving away from the years of hard work within CUSO that went into the formulation of more political support to local partners who aspired to challenge global capitalism, structural adjustment policies and the authoritarian state. The impetus for the regional programme originated in Mozambique and South Africa around support, initially, for their liberation movements. Support was subsequently extended to the labour and co-operative movements.[43] CUSO programmes were then extended to other front line states such as Tanzania. In the late 1980s, programmes began to focus on support for social movements, civil society and land rights, as the original promise of the progressive development projects began to wane, like that of the post-colonial states ruled by FRELIMO (Mozambique), MPLA (Angola), ZANU (Zimbabwe) and CCM (Tanzania). Issues around the empowerment of civil society and social movements for peoples marginalized by state-sponsored structural adjustment policies thus increasingly came to the fore, including those considered to be indigenous peoples.[44]

But by the 1990s, with local CUSO offices shutting down or being downsized, there was a tendency to hold onto local Tanzanian partners and place cooperants uncritically into NGOs like PINGOs. The local CUSO office also resisted calls, from local partners and cooperants alike, to do external evaluations of the Tanzania country office, which increasingly lacked the means and will to safeguard the social goals of CUSO.[45] The Tanzanian placements began unravelling. Former Voluntary Service Overseas (VSO, a British volunteer sending agency known for its frugal budgets) managers were hired by CUSO to downsize the Tanzania country programme in the wake of the ongoing CIDA cuts, in effect an internal structural adjustment policy that witnessed the management requesting voluntary pay cuts among cooperants in line with VSO volunteer salaries.[46] Cooperants were expected to fundraise for their own placements, or do joint fundraising with local partners in order to meet their living costs.[47] The weaker support for cooperants, such as regular and frequent meetings and legal backup for politically sensitive situations, meant that there was little opportunity to share information and offer mutual advice to prevent cooperants from working in isolation from one another and the local CUSO office. The other two cooperants in the Land Rights Programme – one being with the Hadzabe programme with the district government and the other with Illaramatak – had to work hard not to succumb to work tendencies such as gatekeeper roles, or the romanticization of a 'people' and charismatic 'big men'. There were certain essentializations on my part, as well. I initially perceived PINGOs as analogous to a peasant association that should engage with pastoralist issues through rural development work, rather than lobbying and advocating their concerns at the national level. I then swung the other way and accepted indigenous solidarity discourse where, for instance, I bought a map for the PINGOs office of Canada's First Nations, and considered facilitating linkages with a Mi'kmaq reservation in eastern Canada. Eventually it dawned on me that PINGOs was neither a rural development association nor an indigenous apex organization, let alone a lobbying and coordinating Forum. In essence, PINGOs, with all its original potential, had become a Maasai association run by a narrow elite.

The CUSO cutbacks caused much strife and lack of coherence throughout eastern and southern Africa in which CUSO sought to maintain its increasingly shaky programmes. The critical support for the land rights movement in Tanzania eventually fell by the wayside in favour of an emphasis on supporting indigenous human rights, and thus paralleling the approaches of foreign based indigenous rights NGOs. Certainly there was support among some Western donors, including elements inside of CUSO, for more political forms of work, but a subtle shift in emphasis seemed to occur in tandem with the changing logic of their own missions and funding bases in the West. The shift followed, at least in the Canadian case, the trade-not-aid emphasis in official bilateral agencies. CUSO, dependent upon CIDA for most of its funding, underwent severe budget cuts as the Canadian government both shifted the rules of the game, as it were, and also redirected significant portions of the national aid budget to Asia and Eastern Europe post-1989. Thus the taxes of Canadian working people were being squandered due in part to bilateral aid policies in Tanzania, which at the same time gave succour to those reactionary quarters in the government and corporate community seeking to reduce aid budgets and sustainable trade commitments, and economic alternatives. Within this wider aid milieu, therefore, CUSO and its stakeholders in the Land Rights Movement failed to articulate an alternative and socially just vision that questioned the emerging neo-liberal consensus inside and around the PINGOs Forum.

By the mid-1990s, various international indigenous rights NGOs, based in Europe, were increasingly available to facilitate this process of linking the local struggles of pastoralists and hunter–gatherers to the UN level through logistical and financial assistance. Key for PINGOs in this regard were the international NGOs, Survival International (SI) and IWGIA, based in London and Copenhagen, respectively. The SI Land Rights Programme was an instructive example of the mutual benefits accruing to NGO local leaders and donor-initiated programmes that were at best of doubtful benefit to the PINGOs affiliate members. The SI programme initially sought to internationalize the pastoral land struggle and bring about sweeping changes to pastoral land law in East Africa based on precedents in Australian courts, the home country of the SI consultant, who was trained as an anthropologist.[48] The collapse of an initiative to form a local coordinating body of pastoralist experts and activists in Tanzania saw the legal initiative put aside. SI and its consultant, who by then had set up a UK NGO called Pilot Light, reconstituted the programme into a Tanzania/ Australia exchange programme. This initiative involved an extended six-week exchange of Tanzanian pastoralist leaders and professionals, including the PINGOs coordinator, to Australia (with a return trip of Australian Aboriginal leaders and activists to Tanzania at a later date).[49] The consultant envisioned that East African pastoralists and Australian Aborigines would benefit from an open exchange of information and ideas around land issues and indigenous problems common to both East African pastoralists and Australian Aboriginal peoples. Though this exchange would have undoubtedly raised the awareness of pastoral NGO leaders of a geographical corner of the worldwide indigenous rights movement, the impact of the overseas exchange on the forum was nugatory. The coordinator's six-week tour of Australia froze the secretariat since no decisions or fund disbursements could be made without his physical presence on the forum premises. The SI programme also cost thousands of British pounds, money that in an effective Tanzanian institution committed to the PINGOs mission could have been employed in impoverished communities for legal aid training,[50] community leadership promotion, development projects or national level advocacy and lobbying the Tanzanian government.

Even assuming institutional capacity existed in PINGOs to utilize the increased knowledge derived from such international tours, there existed scant will among the beneficiaries to do so. The PINGOs leadership showed little interest in networking with other indigenous peoples abroad unless financed by donors; and even then the beneficiaries of such overseas exchanges made little attempt to diffuse the benefits of these tours to the affiliate PINGOs members. There were no follow-up seminars, packaged to the membership with their own community issues in mind, or even a report produced by the PINGOs leadership for the members. Indeed, how would local activists in the affiliates make sense of such radically different socioeconomic and political contexts? The main comment of a Barabaig activist upon returning from the Australian tour was that the Australian government financially supported Aboriginal organizations citing, for example, community radio stations.[51] Not only were the Tanzanian and Australian judicial systems different from one another, but so was the nature of state and society. To cite another example of this kind of international legalistic reasoning, this SI consultant, during an ad hoc meeting at the PINGOs Forum, made a video presentation on Aboriginal and Maori land cases in Australasia to PINGOs members, followed by discussion.[52] The consultant emphasized that it was important to prove not only occupation of land prior to

European occupation, but that they had their own system of government and cultural norms prior to the arrival of Europeans. The consultant said that through oral history and archaeology the Aboriginal representatives had proven physical and cultural occupation of their land in the Australian courts, a precedent which, he stressed, could be used in Tanzanian courts. The Barabaig representative raised a question on the possibility of compensation, and if forthcoming whether the Barabaig would opt for compensation in land or money.[53] But these were legal remedies underpinned by political and historical processes specific to a capitalist democracy. Could such support for indigenous NGOs be expected in Tanzania?

In contradistinction to overseas exchange tours, the IWGIA-organized conference in Arusha was an effort to bring together indigenous leaders from pastoralist and hunter–gatherer organizations from East, Central and Southern Africa – including the San from Namibia and South Africa, Pygmies from the Democratic Republic of Congo, the Hadzabe from Tanzania, the Ogiek from Kenya, the Himba of Namibia, Somali from Northern Kenya, Maasai from Tanzania and Kenya and the Barabaig from Tanzania – as well as regional and international indigenous networks, IPACC, the Working Group of Indigenous Minorities in Southern Africa (WIMSA) based in Botswana and AIWO (the African Indigenous Women Organization, the indigenous women's apex for Africa),[54] together with NGO activists and specialists from Western countries. The conference discussed the very real problems faced by the various communities represented, discussing for over five days the burning problems, among others, of discrimination, marginalization, land alienation and poor social services suffered by local communities. Reasoning that indigenous self-organization was still in its infancy in Africa, IWGIA was anxious to promote contacts, exchange experiences, and put together forward-looking strategies. On the surface, the conference appeared to represent an initiative based on indigenous knowledge and with a focus on popular participation and planning-from-below.

But IWGIA was more than a facilitator in setting the framework for a discussion on indigenous rights in East Africa. Shortly after the IWGIA conference in August 1998, the main Dutch donor severed the partnership with PINGOs once the damning report from the Novib consultants reached The Hague.[55] As PINGOs became dormant, once funding ceased from Novib, and with its staff scattered, the remnants of the PINGOs leadership, more ambitiously, sought to establish yet another indigenous apex representing the African continent, which they called the Organization of Indigenous Peoples in Africa (OIPA). It is not clear the extent to which IWGIA was involved with the promotion of OIPA. But there had been tensions with the South African-based IPACC which IWGIA felt was controlled by South African whites based in Cape Town.[56] IPACC quite rightly suspected IWGIA of attempting to circumvent IPACC in order that IWGIA consolidate its indigenous credentials in East Africa. This would explain why IWGIA earlier had sought to work with the dysfunctional PINGOs leadership outside of IPACC, and during a stressful external evaluation inside PINGOs, during the above-mentioned Arusha conference (IWGIA itself was being evaluated by its own main external funder during the conference). Sure enough, OIPA was established in the wake of the collapse of PINGOs in September 1999, the venue for its creation being a training session in Arusha under the auspices of PINGOs and the Saami Council, which was supposed to provide experience and a role model.[57] IWGIA most likely had a facilitation role in the birth of OIPA. Underlying this initiative, was a new chance for the PINGOs leadership to get funding, and an opportunity for IWGIA to get around IPACC which itself had well-grounded suspicions of IWGIA agendas from earlier experience in southern Africa. Tensions subsequently arose between the rival apexes. IPACC claimed that OIPA was an East African regional organization, while people connected to OIPA repeated IWGIA's allegation that IPACC was controlled by whites in its Cape Town secretariat.[58] Thus there existed two organizations claiming to represent Africa's indigenous peoples. This meant time, energy and resources being used to iron out differences between organizations sponsored by different donors, with duplication and political spaces to be worked out between them.[59]

How did NGO politics and donor partnerships – underpinned by significant expenditures of time, energy and resources – translate in terms of indigenous knowledge? Working through individuals rather than institutions, donors are able to write proposals as they see fit, advocate in the West with limited accountability, create structures in developing countries when they want to implement their programmes, even leave the locale when a project flounders or a better career opportunity arises elsewhere. Donors can also impose UN discourses uncritically into a local area, ignoring a regional perspective that warns against reification of the 'indigenous'. In the West, it would appear that ever greater numbers of NGOs and Third Sector organizations were competing for ever decreasing sums of public and private money. This required ever more daring fund raising techniques and strategies. In this way, Maasai pastoralist issues afforded donors an opportunity to present the Maasai as a bounded and encapsulated people under threat of losing their way of life in 'vanishing Africa'. IWGIA, for example, was so pleased with its new-found alliance that PINGOs secretariat staff, attired in traditional Maasai dress, graced the cover of its subsequent magazine.[60] In fact, transnational NGOs programming with pastoralist and hunter–gatherers at times appeared to be the flip side of the top-down methodology of conservation NGOs.[61] It was an apolitical representation of the exotic in the West rather than presented as part of a wider subaltern social struggle against both national and international interests; yet another anti-politics machine where both local and international NGO leaders had a vested interest in *not* spawning a land rights movement that could escape their control.[62] Significantly, anthropologists from the field were occupying many of these international indigenous rights NGOs, especially those working specifically on worldwide indigenous rights. Paradoxically in opposing top-down developmentalism of a statist kind, indigenous rights discourses were in fact another form of developmentalism, where pastoralists, hunter–gatherers, and other peoples from afar, were all lumped together for the programmatic convenience of donors. IWGIA and other transnational actors were attempting to glean and filter smaller collectivities of African peoples from larger ones through employing its notion of indigeneity from above. Something not addressed thus far by those sympathetic to international indigenous rights, moreover, is whether transnational NGOs, let alone Western governments, would support Africans deemed indigenous beyond short-term funding cycles or development fashions? More likely, transnational NGOs would sap energy from local efforts at coalition building as well as raising false hopes as to the moral support pastoralist NGOs could expect from the international advocacy NGOs in the medium term.

A communitarian model of society and state

By communitarian view we mean a perspective that conceives of local indigenous communities as cohesive and organic; such a perspective downplays the various power configurations within these communities, while portraying the dominant Third World state as monolithic and hostile to a particular people. The simplistic conception of the state held by the transnational NGOs legitimized their own role and sidestepped the arenas of national and regional politics; it thus allowed intervention despite the lack of local and historical knowledge. Donors' conceptions of the Maasai as indigenous and victimized justified their interventions locally as well as their advocacy role in the West. Yet their world view supported or left unstated the fact of neo-liberal restructuring of the Third World state, the processes of privatization as well as the structural adjustment programmes promoted by Western governments.

Within PINGOs there was never serious discussion around what indigenous meant in Tanzania to Tanzanians and what the implications were for the actualization of the concept in programmatic terms. What did indigenous mean for identity politics in northern Tanzania when a Maasai NGO activist in PINGOs could, for instance, identify with clan, age, gender, class, nation, the East Africa region, or potentially the African continent and beyond? Conversely, did Tanzanians view Maasai, Barabaig and Hadzabe as indigenous or merely behind-the-times? And if the latter, how then could such prejudices against traditional pastoralists and hunter–gatherers be challenged in Tanzanian society? There are, in fact, multiple and contested meanings of indigenous rights in Tanzania. Indigenous (*uzawa*, in Swahili), for example, may refer to the views of those African Tanzanians with the financial capital to aspire to competing with Asian Tanzanian businessmen for business hegemony in the national economy. There also exists indigenous politics in Zanzibar, with those supporters of the CCM regime, or those of African descent, being known as *Wazanzibara* (mainlander Zanzibaris), while those with longer lineages (Persian [Shirazi], Arab or of East African coastal origin), and of a more Zanzibari nationalist bent, being known as *Wazanzibari* (indigenous Zanzibaris). The increasing use of identity labels –

occasionally PINGOs secretariat staff would use the Swahili self-appellation *asili* (original) – would lead to questionable outcomes in wider Tanzanian society if such identity markers over say, land rights, and using such labels as 'the original inhabitants', began to percolate within local communities outside the confines of NGO offices.[63] Even here, *asili* would apply to the Hadzabe, if any one ethnic group, who were the original inhabitants of their area. Woodburn points out that though pastoralists have lost much land, unlike many hunter–gatherers, they are not landless. He goes on to suggest ways in which the positive elements of stereotypes on hunter–gatherers, particularly the value attached to being the first people in an area, could be politically used in the struggle to obtain and retain land, and other crucial citizenship rights. Woodburn feels that a right-oriented approach to politics is growing among Africans, citing the legal victories of Aborigines in Australia who are gradually getting some of their land back. But even in the case of the Hadzabe, Woodburn argues that they should campaign as a First People, and not as indigenous people, since there exists a strong feeling among Africans that, since the end of European colonization, they are all indigenous.[64] However, who would be the active agent in actualizing these legal openings within Tanzania is left unclear, as well as the process involved.

The PINGOs leadership, as we have argued, had already marginalized or alienated non-Maasai affiliates. Neither were real linkages made with other pastoralists let alone hunter–gatherers (except as representations in PINGOs funding proposals to donors). PINGOs leaders and their immediate allies viewed peasant communities monolithically as 'part of the problem' because of encroachment onto pasture lands, despite land alienation and degradation blighting these very same cultivator communities. Such communitarian views on the part of the PINGOs leadership ignored the extent to which existing communities in Africa have themselves been the sites of inequalities and the instigators of exclusions: of women, the young, pastoralists, descendants of slaves.[65] One could add the growing ranks of Maasai men working as exploited night watchmen in urban areas like Arusha and Dar es Salaam.[66] I can only surmise what other communities of peasants, pastoralists and agro-pastoralists, who were actually aware of the PINGOs Forum, thought of it. Certainly transnational indigenous discourse draws our attention to the complexities of *all* local communities – in terms of traditional knowledge, collective rights, natural resource land management and environmental custodianship – and the destructiveness of developmentalism be it of a statist or market kind. However it should not be the handle on which pastoralist NGOs enter local communities to facilitate community responses to developmentalism due to potentially divisive outcomes. An indicator of just such a possibility occurred during a meeting with some of the secretariat staff and a delegation from the Swedish bilateral agency, the Swedish International Development Agency (SIDA).[67] SIDA personnel had come to follow up their support for a community radio station in the area of the coordinator's NGO, Illaramatak. The donor asked if Illaramatak's radio programmes would be Maasai-centric or more inclusive. The PINGOs representative replied that it would be inclusive of 'all indigenous people'. The fact that he said 'indigenous people' had an exclusionary aspect to it, leaving aside who was and was not indigenous in Simanjiro district where Illaramatak was located. It brought to light the possibility that a universalizing indigenous rights discourse articulated, say through community radio (as long as donors were willing to finance it) and under the control of autocratic NGO leaders, could potentially have conservative political implications if those with indigenous knowledge were mystifying power configurations to the advantage of their organizations. Articulating grievances along an ethnic axis would eventually elicit parallel responses among other marginalized communities, and at the same time would most likely make no headway against the regime's policies.

Human rights discourses and the politics of identity have begun to fill a vacuum left by the politics of *ujamaa* (*ujamaa* being Swahili for familyhood or socialism). For the pastoralist NGOs of the PINGOs Forum, the emphasis on international over national institutions for participation, and the defence of their rights, reflected a real lack of participation in state institutions and an awareness of their lowly status in national society, which is now as indicative of indigenous status as it was of peasant status when Wolf defined it in the 1950s.[68] Yet what would identity politics, in fact, mean to the Tanzanian state? Would indigenous politics threaten the state's authoritarian developmentalism? Perhaps the African state might even fear the rights to self-determination as part of the Draft Declaration.[69] Would there be the possibility of further marginalization for pastoralist NGOs? Or perhaps the state would grant such a space to identity politics in order to regulate dissent, meet donors' concern about human rights, and co-opt through divide and rule policies? These were among some of the questions the PINGOs leadership and members had yet to democratically address inside of the forum. This was all too evident when, in June 1999, an IPACC representative met with a group of leaders and members of the PINGOs Forum, in an ad hoc meeting, to explain the UN process and the possibility of sending representatives to Geneva. The coordinator explained the ongoing activities of PINGOs including court cases, lobbying the government via the Land Rights Institute on the land bill, advocacy and civic education, adding that it was good to build a common front because indigenous people are afraid of governments. The coordinator then admitted that he had no idea of how to define indigenous, though he opined that perhaps the international level could aid in this process of self-definition, including the opportunity afforded by donor support for information exchange with other African indigenous peoples. The PINGOs members present in the meeting room were of the view that the international front was more important than lobbying the Tanzanian government, that international law should be a leverage to pressure the Tanzanian government to respect human rights. The coordinator expressed his view succinctly on how the issue of indigenous rights would be received by Tanzanian policy makers: 'we are all Africans why do you call yourself indigenous, we all are', a reference to the general perception of indigeneity as being that of the former relationship between white colonists and the long resident African population.[70] Other PINGOs members wanted to attend the UN meetings but did not know how to go about this; nor were they sure about how IPACC functioned and its responsibility.[71] The IPACC representative pointed out that most African states were wary of special group/ethnic rights, hence the resistance on the part of African states to supporting the UN process which would culminate in a vote in the UN General Assembly at the end of the Decade of Indigenous Peoples.[72] As a strategy to counter this, the IPACC representative felt there might be possibilities for the indigenous peoples of the North to lobby their respective Western governments – particularly the governments of Denmark, Canada, Russia, Sweden and Finland – to in turn put pressure on African governments to support the aims of the UN Draft Declaration.[73] Yet compared to indigenous organizations from other parts of world, like the Americas and Australasia, African organizations, like PINGOs, were potentially entering the global arena of meetings without having had the time to go through the stage of local mobilization, cultural consolidation, capacity building, and above all representativeness. Certainly it is difficult to determine when an organization becomes representative. Observers frequently point out that many African NGOs must use modern structures with the paradox that the more that cosmopolitan English speaking leaders become legitimate internationally, the less they become so locally. Yet in PINGOs, there were more than enough talented activists to institute a division of labour between the local, national and international, had the organization been democratically accountable. Which raises the question unasked by many observers, whose work we have cited, as to when, why and how a hitherto representative popular organization, such as the PINGOs Forum, becomes unrepresentative?

Saugestad argues that talking about cultural differences, identity and contestation of meaning, may appear less radical than arguments along the lines of class difference. Saugestad suggests, however, that a focus on class conflict may appear less controversial to many governments because it simply implies a tacit acceptance of the view that the problem of indigenous peoples is one of poverty only. Yet the underlying mechanisms of unequal opportunity can never be removed by welfare, but require a change in the dominant political discourse.[74] However, as I have argued, indigenous rights could themselves become a dominant political discourse. Saugestad in fact conflates class conflict with welfarism and economism. A less caricaturized radical perspective would go beyond welfarism to instead advocate a bottom-up popular movement, either as a social movement or a political party, or some combination of both, and in alliance with so-called dominant peoples. Turning to the specific context of Tanzania, would the regime actually respond to the initiatives of donors and local NGOs advocating on behalf of Tanzania's indigenous peoples within welfarist social democracy? From Mexico, where popular organizations are much stronger than in Tanzania, and where indigeneity has a real resonance in society because of the Spanish colonial experience, J. Gledhill argues against indigenous peoples being given distinctive rights on the basis of an indigenous identity because victimhood status and culture then become fixed when granted legal status.[75] Moreover an indigenous person is neither transparent nor immune from disputation in any given social and political context: including who is a valid beneficiary, hybridity of cultures, historical connections between past and present peoples, biological purity and so forth. Indigeneity allows certain groups to escape anonymity and achieve a distinct social personality, affording recognition from both national states and the advocacy and support networks of transnational actors, yet may leave out other equally disadvantaged citizens.[76] Raising similar concerns, Tanzanian intellectuals such as I. Shivji argue that the Maasai would become separate and isolated through international indigenous rights advocacy and that, though they have a certain history from Tanzania's development experience, their plight differs little from other citizens of the Tanzanian non-elite.[77]

But why were bilateral and multilateral agencies in Western countries sponsoring their respective national NGOs to support indigenous rights in Tanzania? After all, the majority of these bigger players were supporting macro-level market reforms in Tanzania where one could reasonably make the argument that these policies were causally responsible for exacerbating the very land alienation and 'fragmentation of locality' that indigenous communities were suffering from at the community level in the first place. Many bilateral agencies of Western governments actively implementing structural adjustment policies and other reforms were, at the same time, funding Western NGOs working on indigenous rights in Africa. In Tanzania increasingly, indigenous rights appeared to be a popular sector among bilateral founders like DFID, Danida and CIDA. For instance, the British Department for International Development (DFID) financed a good part of the process behind the pro-investor Land Act, while seeking local partners, and generating a *raison d'etre* for its bureaucracy, to work on Hadzabe hunter–gatherer community issues around land rights.[78] CIDA itself has shown a growing interest in indigenous rights.[79] While in the case of the Nordic countries, there was the solid financial backing offered by DANIDA to IWGIA. Perhaps indigenous rights as manifested in Tanzania potentially weakened the post-*ujamaa* state with the concomitant possibility of an exacerbation of ethnic divisions, while leaving the more general structural-level neo-liberal reforms unchallenged. And here, the international NGOs are able to bypass the Tanzanian state more than ever before to programme directly at the micro-level.

But there are implications for the local level in treading the path of international priorities as contexts elsewhere suggest. In Bolivia, where the majority of people are indigenous native Quechua peoples, the trajectory of identity politics is in flux, and where being both *campesino* (peasant) and indigenous has become contradictory. In Bolivia, the World Bank, international NGOs and the Bolivian state, have sought, consciously or otherwise, to create new identities based on the legal and administrative revival of the *ayllu* (the pre-colonial Andean indigenous model), and marginalize the role of the leftist peasant federations. Bolivian indigenous organizations have in fact forged links to the UN and the Indigenous Working Group. The World Bank has an indigenous development programme and is now interested in indigenous issues.[80] In another part of Latin America, Gledhill notes that indigenous politics in Mexico do contest neo-liberal and bureaucratic regimes by seeking non-individualistic rights of entitlement predicated upon a moral economy. Whether this would lead to a form of left-wing populism and/or millenarianism is highly doubtful in the case of the PINGOs Forum and its affiliates. The agency available to PINGOs saw its leadership uncritically accept the agendas of foreign NGOs, and hence neither a communitarian response nor a coalition building approach became possible.

Conclusion

In the experiences described in this chapter, donors' grand narratives claimed a superior organizational and conceptual knowledge over that of the PINGOs grassroots membership. International perspectives on indigenous rights were employed to transcend all parochialisms in order to mobilize worldwide support for a set of ideals with the hope that local applications would ensue. It is a top-down perspective – the definitions are absolute – where one either has a declared right or one does not.[81] In actualizing this top-down perspective, donors' conception of the pastoral land rights movement sought its mirror reflection in the structures and programmes created through the PINGOs secretariat. This was the basis of the claims to legitimacy by donors who did not look too closely, otherwise, at how the PINGOs leadership conducted its affairs. PINGOs became more concentrated on the policies and funds of donors than they were in struggling to meet the aspirations of their local constituencies. While leaders were co-opted and internal struggles unleashed over donor funds there was little in the way of accountability between the PINGOs leadership and its constituency. This is not particular to pastoralist NGOs, moreover. Manji doubts the ability of Tanzanian feminist NGOs to address the rural land tenure question primarily because of the urban and class biases of these gender NGOs, anchored as they are in Dar es Salaam and very dependent on donor funds. Their constituency is the donor community with the result that Tanzania-specific issues of economic deprivation and unequal access to productive resources by women fall by the wayside;[82] one could almost replace 'feminist' with 'pastoralist'/'indigenous' vis-à-vis Manji's observations. For the PINGOs leadership, such a symbiosis perpetuated their status and patronage networks as they sought to become brokers between their members and the donors, just as the donors themselves had become brokers between PINGOs and their bilateral donors. In the end, the PINGOs leadership purported to represent pastoralists and portray them as one voice behind the educated PINGOs leadership, when the constituency truly served was far narrower. The experience of the PINGOs Forum also illustrates the urgent theoretical work required, both locally and internationally, to make sense of the seemingly contradictory phenomenon of neo-traditionalism's project to preserve what remains of pre-capitalist relations within transnational capitalism's relentless commodification of land and labour relations in non-Western social formations.

Notes

1 For earlier readings of this draft I extend my appreciation to Nayanika Mookherjee and, above all, Justine Lucas whose brilliant intellectual energy during a time of terminal illness was truly inspiring to those of us who had the fortune briefly to know her. I alone remain responsible for the arguments in this chapter.
2 See I. Shivji (1998).
3 The National Food Corporation (NAFCO) was the parastatal that farmed the wheat on the land seized from the Barabaig in the 1970s. For a background on the land conflict between the Barabaig community and the Tanzanian government see Minority Rights Group International (1994).
4 PINGOs Funding Proposal, 1998.
5 PINGOs Funding Proposal, 1998.
6 See the Azimio la Uhai (Declaration of Life) published by the Land Rights Research and Resources Institute [no year].
7 For an examination of the national context within which the PINGOs Forum programmed see G. Cameron (2001).
8 See Pottier (2000).
9 NACLA (1996).
10 PINGOs Funding Proposal, 1998.
11 Minority Rights Group International (1994).
12 Ibid.
13 United Nations Draft Declaration of Indigenous Peoples, Sub-Commission resolution 1994/5. There is also now a Permanent Forum on Indigenous Issues, the first session being held in May 2002. Office of the High Commissioner for Human Rights, www.unhchr.ch.
14 Wilson (1997: 10).
15 Thornbury (2000: 1).
16 For the complex legal distinctions between 'indigenous' and 'minority' see Thornbury (2000: 5–6).
17 Thornbury also makes the point that, in addition to indigenous rights law, indigenous people can make claims on general human rights international law as indigenous people (2000: 8). There is also a fuller theoretical treatment of general human rights discourses in Wilson (1997).
18 IWGIA (1999).
19 Ibid.
20 Saugestad (2000: 8).
21 IPACC (1997). A Danish NGO, based in Tanzania for many years, and which was very close to the Illaramatak leadership (though it didn't programme directly with PINGOs), saw pastoralists and hunter–gatherers as having four issues in common that fitted the concept of indigenous and hence required donor support: government policies negatively impacting on their ability to maintain traditional production systems; production systems based on the use of common-pool resources; the practice of the communal management of common-pool resources; and livelihoods which

depend on a close relation between their production system, culture and natural resources. Danish MS preferred the terms of *wafugaji* and *wakusanja* (pastoralists and hunter–gatherers, in Swahili, respectively) while working in Tanzania since these were the emic self-appellations of these peoples. Internationally, however, Danish MS used the term indigenous. The document does recognize, at least in theory, that there exist gender and class inequalities in these communities MS (1994: 10, 18).

22 IPACC has membership from 11 of these, with the highest coverage in East Africa and the lowest in West Africa (1998b).

23 NGOs outside of the donor/PINGOs circuit, for instance, had not employed indigenous rights discourse. For instance, Inyuat e Maa failed to gain admittance to PINGOs due to an earlier struggle with the PINGOs coordinator. Though it showed a particular Maa culturalist bent, Inyuat e Maa had not, at this time at least, meshed this culturalist bent with international indigenous law. This may have been because of a brief abandonment by donors, and a later partnership with a conservation donor that would have had little truck with the concept (Inyuat e Maa 1997). Another NGO, while ethnically oriented to Maa speakers (the agriculturalist Waarusha), either did not know of indigenous rights, or perhaps was too close to government officials to employ the term in their funding proposals. (Mukulat Advancement Association 1997).

24 At one meeting with PINGOs members (9 June 1998), an IPACC representative said that the European Union wanted to help but did not yet know who the indigenous were in the African context. This reference to the EU by the IPACC representative may have referred to the EU draft policy on support for indigenous peoples. This EU draft policy document attempted to identify such partners and provide ethical guidelines to national EU bilateral institutions, such as the British Department for International Development (DFID) (EU 1995).

25 Woodburn (2000: 8).

26 PINGOs correspondence to IWGIA, 24 March 1997.

27 Ibid.

28 The Akie and Aramanik (generally referred to as the Dorobo) are a hunting and gathering people inhabiting forests that are interspersed among woodlands and plains inhabited mainly by the pastoral Maasai (Kaare 1998: 4).

29 PINGOs files.

30 PINGOs files.

31 KIPOC Barabaig to the International Institute for the Environment and Development (IIED), dated March 30 1993.

32 Inyuat e Moipo Funding Proposal (1998).

33 (Bulgalda 1997); personal communication with Barabaig leader, July 1996. When the Barabaig leaders resigned, the PINGOs members, upset at their withdrawal from PINGOs, wondered what the Barabaig leaders were up to. PINGOs minutes of 3–8 July 1996 General Meeting; personal observation. Bulgalda later applied to rejoin when it changed leadership.

34 The newsletter front page also read in Swahili: *Jarida la Maendeleo ya Wamasai* (The Newsletter of the Maasai). Joint Committee Meeting, 22 April 1997.

35 Kaare (1998: 27).

36 There were ongoing efforts by the leadership to seek funds from UN bodies like UNESCO to enable PINGOs to set up and run a culturally appropriate pastoralist college, an indication of further bureaucratic aspirations by the PINGOs leadership.

37 Aside from pastoralist NGOs in Tanzania, such governance problems were also to be found in Kenya in local government, community conservation programmes, and NGOs (Homewood 2001).

38 Novib employs about 250 people and is active in more than 45 countries. Most of its income derives from the EU and the Dutch government. Novib is connected to

Oxfam International and does work in the North via policies seen as detrimental to the Global South such as agricultural subsidies. Novib appeared to be more focused on traditional development work, like sustainable development projects, and human rights more broadly conceived, such as governance issues around citizenship rights, rather than on international indigenous human rights (Novib brochure).

39 For a comparative analysis of marginalized first peoples in Australia and Canada see Young (1995).

40 I joined PINGOs Forum on a full-time basis in 1997 and worked as an administrator and training staff person until the end of 1998 when I resigned because of mismanagement in the PINGOs secretariat.

41 Union of Ontario Indians (1995).

42 Union of Ontario Indians (1995: 14). The PINGOs coordinator was well aware of the Canadian native struggles, perhaps from the experiences of these earlier delegates, requesting from a CUSO officer strategic advice and information from the 'Indian struggles in Canada' (Secretariat Meeting, 17 February 1998).

43 Internal email correspondence, 24 April 1999.

44 CUSO did not have an official position paper on indigenous peoples in Tanzania. Its approach, rather, was to assess a particular community organization's proposal for partnership in the light of CUSO's regional and national missions outlined above. A CUSO regional staff person, during a courtesy call to PINGOs headquarters, explained this to PINGOs secretariat staff who had expected such a policy document, perhaps in line with Danish MS (Secretariat Staff Meeting, 17 February 1997).

45 CUSO Regional Meeting minutes, 1998.

46 Due to lack of funds for transport in the wake of the CIDA cuts, the newly-hired CUSO Tanzania field staff officer (and former VSO manager) wanted to redirect future cooperant placements in the urban sector or in the posting of CUSO teachers 'to support schools and adult education projects which benefit pastoralists and hunter–gatherers?' Such top-down and state-centric placements had been discredited and abandoned some years earlier in the CUSO region due to their overall ineffectiveness in the light of structural adjustment policies (SAP). In another VSO-like recommendation, this programme director asked cooperants to self-SAP themselves by cutting their basic subsidies: 'It seems to me to contradict the notion that cooperants should live frugally and experience a quality of life as close as possible to that of their Tanzanian colleagues' (Correspondence from CUSO Tanzania field staff officer to CUSO cooperants, 22 July 1997). Years earlier, while working on Zanzibar, I had met this current CUSO Tanzania manager (1997) when he was the VSO Tanzania manager (1989). I was present when, during an impromptu meeting with VSO volunteers on the island of Pemba, this manager was warding off complaints from VSO volunteers at their extremely low levels of support from the VSO office, such as subsidies and professional support, as compared to other national volunteer sending agencies (like CUSO or Nordic and Dutch agencies such as SNV). The VSO volunteers recognized, unlike their VSO manager, that to be 'closer to the people and local colleagues' in a meaningful and effective way required adequate professional support, and that such support was a very important consideration to local Tanzanian institutions (e.g. transport) in requesting VSO placements in the first place.

47 Correspondence from CUSO programme officer to cooperants, 17 February 1997. In 1999, CUSO cooperants tried to get explicit support from the CUSO centre to incorporate the African indigenous movement, as part of the Decade of Indigenous Peoples, into a wider vision of global activism. It is unclear if capacity existed within CUSO to implement such a resolution (CUSO email correspondence, June 1999).

48 Igoe (2000: 12). See also Survival International (1998).

49 For details of the Australia tour see www.whoseland.com.

50 Oxfam had become involved with the PINGOs leadership, running a para-legal training programme that led to the usual problems of financial transparency and flawed implementation at the local level. It is not clear to what extent Oxfam itself was gravitating to an indigenous rights programme and away from its earlier solid work with the Land Rights campaign at the national level in Tanzania.
51 Personal discussion, 4 May 1998.
52 PINGOs Forum meeting, 14 August 1996.
53 CIDA attempted to compensate the general population of Hanang District, where the wheat farms were set up, with a Community Development Fund. However, CIDA maintained a hands-off approach via the question of a return of the expropriated land, numbering in the thousands of acres of pasture and water land, to the Barabaig people who to the present time had not been compensated.
54 IWGIA (1999).
55 A. Umar and E. Yamet (1999).
56 Personal communication, 19 August 1998.
57 Saugestad (2000: 20).
58 Ibid.
59 Saugestad (2000: 20–21). Saugestad is somewhat sanguine over the role of IWGIA in creating this divisive state of affairs in the first place. OIPA never took off, and an organic alternative, unencumbered by donors, has yet to emerge in Tanzania. Personal communication 27 June 2002.
60 IWGIA (1999). IWGIA planned to hold a similar conference in West Africa.
61 The Tanzanian government evicted Maasai and Pare peoples who had long resided within a customary land tenure regime on the reserve. Controversially, the move was supported by a conservation organization, in the preservationist tradition, known as the Tony Fitzjohn/George Adamson African Wildlife Preservation Trust. This conservation outfit backed the draconian expulsions by the government, regarding the evictions of the pastoralists and their livestock as necessary for the conservation of the wildlife in the reserve. See Mustafa (1997).
62 An extreme example of this romanticized phenomenon is the radical naturalist group Friends of Peoples Close to Nature (FPCN) which aims to establish Hadzabe lands as protected territories in order to preserve their culture. In March/April 1997 FPCN sent radical naturalist youth from the UK to 'fact find' on the Hadzabe. Devoid of a grounded local or political perspective in Tanzania, let alone a feasible alternative approach, these youth and their leader would write bombastic reports against other NGOs and local officials upon return to the West. See the FPCN website: www.fpcn-global.org.
63 In standard Swahili dictionaries, and in common everyday use, *asili* commonly means origin, natural source, ancestors, original inhabitants or aborigines (e.g. *watu wa asili*) (Rechbenbach 1968).
64 Woodburn (2000: 8).
65 Moore (1999: 44).
66 An Arusha newspaper published an article on how the Maasai had replaced the Makonde as the night watchmen of choice in 1997.
67 PINGOs meeting, 8 September 1997.
68 Wilson (1997: 10).
69 Article 3 of the UN Draft Declaration.
70 PINGOs meeting, 9 June 1998.
71 IPACC (1998a).
72 As of May 2002, it is still unclear whether the Declaration of Indigenous Peoples would be adopted within the Decade for Indigenous Peoples, ending in 2004. Only two out of around 150 articles of the draft have been agreed on, though delegates participating in the process pointed out that the process itself is positive because it

enhances dialogue between state and peoples around their collective rights (Office of the High Commissioner for Human Rights, www.unhchr.ch.

73 IPACC (1998b: 5).
74 Saugestad (2000: 9).
75 Gledhill (1997: 20).
76 Gledhill (2001: 3–4).
77 I. Shivji and W. Kapinga (1997).
78 DFID letter from the First Secretary (Development) dated July 30, 1998 and titled: British Development Co-operation in Tanzania: 'Partnership With Civil Society', to The Country Representative, CUSO-Tanzania, Arusha; and email correspondence to CUSO Tanzania from a DFID representative, the latter correspondence stating that DFID is 'concerned about the implications of the proposed land bill for hunting and gathering groups in Tanzania', and stressed the importance of consultations around the proposed land bill and the narrow chance for the concerns of hunter–gatherer groups to be included in the consultation process around land reform.
79 An indicator of this interest being the participation of a CIDA representative at the ASA 2000 conference itself from which this volume derives (3–5 April 2000). Frequently citing the burgeoning funds available from CIDA, many of his interventions were directed to the anthropologists present and the possibility of their accessing such funds for research in the area of indigenous knowledge.
80 Androlina (2001). A recent email from the World Bank was circulated to pastoralist activists in Tanzania inviting them to a workshop, perhaps indicating its fledgling interest in indigenous rights in Africa now (Personal communication, March 2002).
81 Moore (1999: 44).
82 Manji (1998: 665).

References

Androlina, R. 2001. Legitimacy and identity in transnational politics. SOAS Department of Politics Seminar Series, 14 March.

Bulgalda. 1997. Report on activities. From April 1996 to February 1997.

Cameron, G. 2001. Taking stock of pastoralist NGOs in Tanzania. In *Review of African Political Economy* 28 (87) (March): 55–72.

CUSO email correspondence, miscellaneous.

CUSO Regional Meeting minutes. 1990–1992, 1997, Zimbabwe and Zambia.

European Union. 1995. Indigenous peoples participation in European Union development policies.

Gledhill, J. 2001. Rights and the poor. ASA conference on Anthropological Perspectives on Rights, Claims and Entitlements. University of Sussex, 30 March–2 April.

Gledhill, J. 1997. Liberalism, socio-economic rights and the politics of identity: From moral economy to indigenous rights. In *Human Rights, Culture and Context: Anthropological perspectives.* (ed.) R. Wilson. London: Pluto Press.

Homewood, K. 2001. Elites, entrepreneurs, and exclusion in Maasailand. SOAS Department of Anthropology Development Studies Seminar Series, 6 March.

Igoe, J. 2000. Scaling up civil society: Donor money, NGOs, and Tanzania's Pastoral Land Tenure Movement. Seminar Paper (no place provided).

Indigenous Peoples of Africa Coordinating Committee (IPACC). 1998. Report from PINGOs visit, 9 June.

Indigenous Peoples of Africa Coordinating Committee (IPACC). 1998. Half Year Report to Breadline Africa, May.

Indigenous Peoples of Africa Coordinating Committee (IPACC). 1997. Minutes, First IPACC Executive Meeting, Cape Town, South Africa, 21–24 November.

International Working Group on Indigenous Affairs (IWGIA). 1999. Indigenous Affairs, No. 2, April–May–June.

Inyuat e Maa. 1997. Position Paper.

Inyuat e Moipo. 1998. Funding Proposal.

Kaare, B. 1998. Report on the land reform process and its implications on the indigenous hunter-gatherer peoples in Tanzania. Report for DFID.

KIPOC-Barabaig correspondence, 1993.

Land Rights Research and Resources Institute [no year]. Azimio la Uhai (The Declaration of Life), Dar es Salaam.

Manji, A. 1998. Gender and politics of the land reform process in Tanzania. In *Journal of Modern African Studies* 36 (4): 646–667.

Minority Rights Group International. 1994. Land rights and minorities profile. July.

Moore, S.F. 1999. Changing African land tenure: reflections on the incapacities of the state. In *Development and Rights: Negotiating justice in changing societies.* (ed.) C. Lund. London: Frank Cass.

Mukulat Advancement Association (MAA). 1997. Strategic plan for ten years from 1997 to 2007.

MS Danish Association for International Co-Operation. 1994. Strategy and action plan for support to pastoralists and hunter gatherers.

Mustafa, K. 1997. *Eviction of pastoralists from the Mkomazi Game Reserve in Tanzania: An historical review.* IIED Drylands Series.

NACLA. 1996. Gaining Ground: The indigenous movement in Latin America (editorial). In *NACLA Report on the Americas.* Volume XXIX, No. 5, March/April.

Novib. brochure (no date or publishing place).

Office of the High Commissioner for Human Rights. May 2002 www.unhchr.ch.

PINGOs Correspondence to IWGIA, 24 March 1997.

PINGOs Minutes of General Meetings, 1997–1999.

PINGOs Funding Proposal, 1998.

Pottier, J. 2000. Modern information warfare versus empirical knowledge: Framing the crisis in Eastern Zaire. ASA Conference on Participating in Development: Approaches to Indigenous Knowledge. SOAS, University of London, 3–5 April.

Rechbenbach, C. W. 1968. *Swahili–English Dictionary.* The Catholic University of America Press, [no place of publication].

Saugestad, S. 2000. Contested images. Conference on Africa's indigenous peoples: First peoples or marginalized minorities? Centre of African Studies, University of Edinburgh, 24–25 May.

Shivji, I. 1998. *Not yet Democracy: Reforming land tenure in Tanzania.* London: IIED.

Shivji, I. and W Kapinga. 1997. Are international groups helping the Maasai? *Arusha Times*, 1–15 November.

Survival International Document. 1998. No future without land.

Thornbury, P. 2000. Indigenous peoples in international law. Conference on Africa's indigenous peoples: first peoples or marginalized minorities? Centre of African Studies, University of Edinburgh, 24–25 May.

Umar, A. and E. Yamet. 1999. Report of the evaluation for Novib of the pastoralist indigenous non-governmental organizations (PINGOs) network in Tanzania.

Union of Ontario Indians. 1995. Re-building the aboriginal economy. 9–11 May.

United Nations. 1994/1995. Draft Declaration on the Rights of Indigenous Peoples, Sub-Commission resolution.

Wilson, R. 1997. Human rights, culture and context: An introduction. In *Human Rights, Culture and Context: Anthropological perspectives.* (ed.) R. Wilson. London: Pluto Press.

Woodburn, J. 2000. The political status of hunter-gatherers in present-day and future Africa. Conference on Africa's indigenous peoples: first peoples or marginalized minorities? Centre of African Studies, University of Edinburgh, 24–25 May.

Young, E. 1995. *Third World in the First. Development and indigenous peoples.* London: Routledge.

Chapter 9

Domestic animal diversity, local knowledge and stockraiser rights

Ilse Köhler-Rollefson and Constance McCorkle

Domestic animal diversity (DAD) is rightly labelled one of the most threatened aspects of biodiversity by the Food and Agriculture Organization (FAO), the UN entity charged with global oversight of DAD documentation and conservation. Much of today's remaining diversity in domestic animal breeds survives in traditional farming and herding communities in the South, where it was generated by local/indigenous knowledge and social organization. Yet FAO and other international organizations have made little effort to integrate such knowledge and practice into their global strategies for understanding and maintaining DAD. Their rationale for saving local/indigenous breeds from impending extinction seems to lie mainly in these animals' possession of valuable genetic material that may be of potential benefit to the North or to humanity at large. (Ironically, many indigenous breeds are now at risk due to cross-breeding policies previously promoted by the same formal-sector institutions now seeking to save them.) Scant attention has been paid to the fact that endangered breeds are frequently associated with marginalized social groups whose economic and cultural survival depends directly upon these animals – and who thus have an even stronger and more immediate interest in their conservation.

Agro-biodiversity is composed of crop and livestock diversity at the levels of genes, species and habitats. In general parlance, however, farm-animal diversity refers to the recognized breeds of ruminants, swine, equines and poultry that humans have developed across the millennia from less than a score of wild animal species. This DAD is now at grave risk. According to the Food and Agriculture Organization, only some 5,000 breeds or strains of all species of farm-animal domesticates still exist, from among untold thousands developed historically. Europe alone has lost nearly half the breeds of farm animals found there at the beginning of the twentieth century. Worldwide, about a third of the remaining 5,000 are in danger of extinction, with breeds disappearing at the rate of more than one per week (Scherf 1996).

In the North, most livestock is raised under intensive or even industrialized production systems that make for impressive outputs of meat, milk, eggs and fibre. These outputs have been achieved by using such high-tech reproductive operations as artificial insemination and embryo transplantation, selecting solely for productivity without regard to other traits such as fitness. In the process, the genetic base of the species so treated has been greatly narrowed and their natural survival traits have been compromised. The resulting food-producing 'machines' are very vulnerable to disease and generally less hardy. They therefore require high veterinary and other inputs, such as special feeds, expensive housing and sophisticated husbandry. Further, these industrial breeds' fertility and reproductive performance have been undermined to such an extent that they often have difficulty mating, giving birth and mothering their young. It is questionable whether they could even survive outside the managed environments and factory farms in which they are raised.

To ensure at least a modicum of fitness and vitality in future populations of food-producing animals, and to keep genetic options open, access to fresh genetic material will always be required. But most of the wild relatives of today's animal domesticates are extinct. This means that a major source of such material lies in the more rustic livestock kept by farmers and herders under extensive, subsistence-oriented production systems in the South.

Variously termed local/indigenous or unimproved (that is, by Northerners), these Southern landraces of animals still harbour such invaluable characteristics as natural disease-resistance, strength, hardiness, good libido and fertility, and other survival-oriented traits (see next section). Indeed, in the absence of livestock with these traits, the majority of the world's people could not survive. Most of the world population is still rural, and all rural groups rely on domesticated animals for a key part of their livelihood; but few can afford or access the high-tech inputs and breeds typical in the North. Fortunately for people everywhere, however, Southerners' landraces often embody special traits that are of present or future economic importance. Their genetic makeup may offer possibilities for tastier, healthier or more specialized foods than the bland, generic products from factory animals.

But if, like their wild relatives, indigenous livestock breeds also disappear, then future human generations will have no genetic manoeuvring room in which to adapt their food-animal production systems to emerging disease threats or changing environmental and economic conditions. Thus serious DAD conservation efforts by FAO and other formal-sector institutions like the International Livestock Research Institute (ILRI) are vital to future food security.

Article 8 of the UN Convention on Biological Diversity states that genetic resources should be conserved in the 'surroundings where they have developed their distinct properties'. For livestock, this means the pastoral and farming communities that nurture most of the world's existing DAD. The diversification of a livestock species into many different strains or breeds is the outcome of different ethnic and social groups' managing that species in a wide variety of habitats and manipulating its genetic composition according to localized biophysical (climate, disease and predator threats, availability of forage, water, minerals, etc.) and sociocultural (e.g. knowledge, beliefs, labour, social organization) conditions in the context of these human groups' varying needs and preferences for livestock goods and services.

In other words, albeit without the North's high tech, people in the South and in traditional societies everywhere have also consciously shaped breeds to their own ends – contrary to ethnocentric beliefs among many animal scientists that livestock in developing countries evolved without human intervention (e.g. Timon 1993). Although targeted studies of animal breeding in traditional cultures are rare, recent overviews of local/indigenous knowledge of and practices in animal reproduction, breeding and other husbandry arenas attest to a wide range of astute interventions in livestock genetics (e.g. Köhler-Rollefson 1997 and 2000; Martin *et al.* 2001; McCorkle 1999). Given their direct dependence on livestock for daily survival, nomadic pastoralists in particular have accumulated a rich store of knowledge and expertise in animal breeding.

Indigenous management of animal genetic resources

Knowledge of breeding. All long-time stockraising peoples have a practical, working knowledge of genetics, at least at a phenotypic level. Many pastoral and agropastoral peoples also keep detailed mental or oral stock records. Indeed, even children can often identify the pedigrees of all the animals in their own and nearby families' herds. This is because an animal's ancestry is often encoded in its name, and names are not changed when animals are sold or exchanged. Usually, records follow female bloodlines.

Furthermore, many pastoral peoples restrict or even taboo the sale of (especially female) breeding stock outside their home community or ethnic segment (tribe, caste). They may do so because they consider such animals their prime capital, as do practically all camel pastoralists (Köhler-Rollefson 1993); or ideological explanations may be offered. For example, Andean agropastoralists' believe that the souls of animals slaughtered far from their home community will be unable to make their way back to be re-born, thus prejudicing continued herd reproduction (McCorkle 1983). Whatever the emic reason, etically such social restrictions isolate animal populations, and thus encourage and stabilize the development of unique, localized breeds.

Breeding goals. Nearly every traditionally stock raising society has developed one or more distinctive livestock breeds to suit its particular environment and animal-product needs and wants. People often have multiple breeding goals for a given species. Still, the first consideration is sheer survivability. Thus stock are selected and bred for: resistance to dangerous endemic diseases; general hardiness, as in an ability to survive even when forage, water and minerals are scarce and to trek long distances in search of them; and adaptability to local climatic extremes, whether in temperature, precipitation or altitude. Also often figuring in breeding choices is animal behaviour and temperament as these relate to livestock survival and human management thereof, for example, good mothering instincts and aggressiveness towards predators, but also herdability, tameness, and loyalty, the latter reflected in stock finding their own way home and resisting contact with humans other than their masters/mistresses.

Of course, in the absence of high-tech operations, good libido and natural fertility are general desiderata. Other breeding goals may involve the particular goods and services sought from a given species or animal, beyond just maximum product yields. To take a few examples: strength or fleetness in draught, pack or riding animals; intelligence in lead or work animals; and particular fibre colours or qualities, milk-fat content or eggshell hardness, according to market demands and conditions. Finally, various aesthetic and cultural considerations may also be at work, as when a people prize certain phenotypic characteristics (e.g. a particular horn or body conformation, feather configuration or physiognomic feature) for special social or ceremonial purposes.

Breeding practices

These consist of the techniques and social institutions by which breeding goals are achieved. The most common is simple selection of animals allowed to mate. As a rule of thumb, traditional stockraisers focus selection on male animals, although typically they also take into account the male's female forebears and relatives.

Selection is negatively implemented by such techniques as: castration – probably the most common way of preventing unsuitable male animals from breeding; culling, slaughtering or selling animals considered unfit as breeding stock; subdividing herds by sex and age; tying aprons or sheaths over the genitalia; and in one culture, inserting stones into the vagina as a sort of IUD. It is positively implemented by, for example, corralling together or manually coupling prime mating pairs; purchasing, hiring or borrowing preferred studs; 'stealth' techniques such as grazing or loosing ones' female animals in another's herd that has desirable studs; and other social mechanisms such as multi-household stud ownership.

The practice of offspring testing is reported for camel breeders in Somalia and India. Also widely used in scientific breeding, offspring testing refers to restricting a male animal's mating until his first crop of progeny can be evaluated. Only if his first offspring live up to expectations will the male then be used more widely.

Some stockraising societies fastidiously avoid inbreeding, while others see no harm in it. Attitudes about animal inbreeding sometimes mirror local marriage practices. For instance Arabian Bedouin are endogamous and so permit inbreeding, whereas the exogamous pastoralists of Rajasthan never allow it. Indeed, breeding practices are often intertwined with cultural beliefs and local social organization. To take another example, the historically superior quality of cattle in the Marwar region of Rajasthan has been attributed to the fact that all sub-standard bulls were systematically castrated (Kothari, personal communication). This practice hinged on the availability of low-caste people to perform this socially debasing task. Nowadays, however, few such people are willing to do this work, and cattle quality has thus suffered (Alstrom 1999).

Breeds. Many local/indigenous livestock species and breeds have a close association with particular ethnic groups with whom they have co-evolved. Indeed, these groups may couch their very identity in terms of the human/animal bond. The survival of the two – animals and humans – is interlinked. Without special interventions, animals will seldom retain their genetic uniqueness outside the traditional production systems that gave them rise. Conversely, the ethnic group in question cannot continue their production system and thus their cultural (or even physical) survival without these animals. A few examples follow, drawn just from among stockraising peoples of a single country, India.

- One of the best indigenous cattle breeds is the Sahiwal, which combines exceptional hardiness with high milk yields. Muslim pastoralists are the hereditary keepers of Sahiwal, which they carefully maintain as purebreds by vigourously resisting all attempts by government and development agencies to introduce cross-breeding (Tantia *et al.* 1998).
- Rebari pastoralists developed the Gir and Kankrej cattle breeds, which became the founding stock of famous beef cattle breeds in Brazil and the southern US But their original creators are being pushed off their pasturelands, thereby imperilling these breeds' continued existence in their original surroundings.
- A hereditary camel-breeding caste, the Raika can no longer earn a living from their breeding work because alienation of pastures has reduced the sizes and reproductive rates of their herds. But regarding themselves as guardians of the camel, some of them continue to keep camels anyway.
- Bastar tribals in Madhya Pradesh are the only group to have protected the rare Aseel chicken, which is extinct everywhere else.

Such examples could be multiplied many times over across the globe (see, for example, ITDG 1996; McCorkle 1999; Vásquez 1997). The point is that the human groups who have nurtured and developed special animal genetic resources should be regarded as their rightful owners. As such, these peoples can play, and should be accorded, a vital key role in initiatives to conserve these invaluable resources.

Conserving livestock diversity and stockraiser rights

FAO has a global mandate to study, advise and set guidelines on conserving livestock genetic resources for present and future food security. FAO's Initiative for Domestic Animal Diversity began by establishing a database known as the DAD Information System (DAD-IS). The system is designed to inventory and monitor DAD worldwide. The Initiative further seeks to conserve those breeds classified in the database as endangered and critical, and to promote exchange of these precious genetic resources. To these ends, the Initiative has established: an intergovernmental mechanism, a technical programme of management support for UN member nations, a cadre of experts, and a country-based global infrastructure of national coordinators.

On paper, FAO recognizes the importance of stockraisers' local/indigenous breeding knowledge. 'The indigenous knowledge that has helped to produce and maintain domestic animal diversity is largely unexplored and yet this knowledge is essential in order to understand and continue developing these animal genetic resources' (FAO n.d.). This recognition has not been translated into any specific activities, however.

Databasing. So far, no efforts have been made to recover local/indigenous knowledge about breeds and then integrate it into the DAD-IS. Database information is provided only by designated national coordinators in FAO member countries. Stockraisers are not directly consulted, even though they may be the only ones who know of many local breeds and strains. To illustrate, only recently did scientists discover a new camel breed with high milk-production potential – although naturally the pastoralists from whom scientists learned of the breed had long known about it (Köhler-Rollefson and Rathore 1995).

Neither does the DAD-IS take account of stockraisers' breed classification systems. These typically differ from scientists' and often are more refined than scientific classifications. To take just one example, scientists opine that India's donkey population is comprised of a single breed. But local donkey-breeders distinguish at least three types that are phenotypically distinct and hail from three different areas. By scientific criteria, these features make them, in all probability, three breeds (or at the very least, three strains).

The DAD-IS characterizes breeds according to production features and population size. The former include: milk yield, lactation length, milk fat, litter size, birth weight, adult weight and adult wither height. But stockraisers evaluate breeds differently, often using non-productivity-related characteristics. (Recall the discussion of breeding goals above.) Many such traits are manifest only under traditional management; they may be invisible in stabled animals or on government research farms. For example, transhumant Gujjars of the Himalayan foothills keep a buffalo breed that instinctively knows when to migrate to higher or lower pastures (Hussain *et al.* 1999). But this behaviour would be unobservable in Gujjar buffalo penned on a farm or research station.

Population data recorded in DAD-IS include, for example, total population size, total number of females bred, total number of males used for breeding, and so forth. Again, this information does not systematically draw on stock-raiser knowledge, breed classifications and terminologies. Consider the case of Tharparkar cattle. Indigenous to the Thar Desert along the Indo-Pakistani border, Tharparkar are famous for their good yields of high-fat milk even under arid conditions. Thus they are accorded a high conservation priority. Yet the word Tharparkar means nothing to local people, who instead say Sindhan. Scientists themselves cannot agree on how to define and hence enumerate this breed. Some count the entire cattle population (several tens of thousands of head) in the two districts of India where it occurs; others consider only the couple of hundred animals kept on state breeding farms to be 'true' Tharparkar.

Conservation. Traditional breeds are most often lost because grazing grounds are appropriated for cropping, wildlife, tourism/recreation, damming, etc.; animal numbers are decimated by both natural (droughts, floods, epidemics) and anthropogenic disasters (famines, wars); technology replaces the services livestock previously provided (e.g. motorized for animal power); inappropriate development policies promote crossing or substitution with exotic breeds; and government policies prejudice pastoral lifeways.

As it now stands, FAO technical information might help *preserve* breeds on government farms. But this is an unsatisfactory solution, whether from the viewpoint of Article 8 or of the people who depend upon the breeds in question. In order to maintain the traits for which they are valued, local/indigenous breeds need to be *conserved* under the rigours of the extensive systems in which they were bred. Once animals are removed from their native environment/management or biocultural regimes, different selection pressures take over. Thus, the focus of conservation should be on rescuing breeds in the surroundings where they have developed their distinct properties, and directly involving the peoples who still husband them there.

Stockraiser rights. Currently, formal-sector institutions give little credit to stockrasing peoples for the crucial role they have played, and continue to play, in DAD conservation. For animal genetic resources, there is as yet no movement paralleling that for crop-farmers' rights. Somehow, intellectual property rights have been seen as less urgent for animal than for plant domesticates. Yet the danger is the same. Genes from indigenous breeds are being appropriated and patented with no concern for stockraiser rights.

For instance, according to some reports, the patented Booroola gene that triggers high multiple-birth rates in Australian Merino sheep originated from

India's native Garole sheep. To take a more recent example, ILRI is working to identify the gene sequence responsible for worm resistance in Red Maasai sheep. This trait is of enormous interest to producers in Australia and other Northern countries where sheep no longer respond to commercial de-worming drugs. But there has been little discussion of how Maasai pastoralists should be recompensed for this intellectual property.

Big corporations are also interested in indigenous animal genetic resources. Their free-for-all bioprospecting for such resources is no longer just the stuff of Hollywood horror movies. As a recent paper on swine genetics recounts:

> Some genotypes formerly not among the ones of economic interest for the [livestock] industry became targets of the breeding companies' research programmes which aimed at discovering and transferring specific genes from these genotypes to the industrial genetic lines. This is, for example, the case with the highly prolific Chinese breeds and the Iberian pig with excellent meat quality for production of extensively cured pork products.
>
> (Pereira *et al.* 1998)

Given that the Northern breeding industry jealously guards and patents its own genetic materials, there is a moral imperative to extend similar protections to traditional stockraisers and breeders – although, granted, this will be no easy task.

Conclusion

It appears that FAO and other publicly funded international institutions concerned with DAD are pursuing their agendas predominantly from the so-called genetic resource angle. Breeds are to be rescued in their role as carriers of genetic material that might have some future value for the livestock industry or humanity at large. Although FAO explicitly recognizes the concept of breed as a cultural rather than a technical construct, there have been no efforts to document and database local knowledge of breeds, nor which ethnic group(s) are closely associated or have co-evolved with different breeds. Likewise for information on the local/indigenous institutions, breeding practices and cultures of the peoples who have nurtured and shaped so many hardy livestock strains. Moreover, there have been no moves to decentralize activities so that these stockraisers can themselves participate in on-the-ground conservation or to ensure that their traditional societies benefit from sharing the unique genetic resources they have created and conserved.

Genetic resources are under the sovereignty of national governments, and legally FAO can work only with government agencies. But ironically, usually it is government programmes and policies that are responsible for massive declines in animal landraces via, for example; government extension services' promotion of crossbreds and exotics; government veterinary services' eradication of whole breeds outright, in the name of disease control; government laws (and corruption) that privilege other land-use options for indigenous breeds' traditional habitats; and even purposive government persecution and ultimately ethnocide of, especially, mobile pastoralists.

Breeds are the product of specific ethnic groups and societies living in specific locales. These peoples often also have a culturally highly developed sense of guardianship, partnership, or even personhood vis-à-vis their animals. This heritage should make them the lead actors in all DAD conservation efforts. It is both technically and ethically imperative to open channels of communication with stockraisers and to institute mechanisms for involving grassroots groups – those who have so wisely shaped and stewarded different breeds down through the centuries and who stand to lose the most if these unique resources disappear from the face of the earth.

In order to do so, it will probably be necessary to involve PVOs and NGOs as mediators (Köhler-Rollefson and Bräunig 1999). It is important to note that almost all successful initiatives to conserve threatened native and 'antique' breeds of the North are due to the enthusiasm of stockraiser and breeder associations and charitable groups, with virtually no government support. A case in point is Germany's Society for the Conservation of Threatened Livestock Breeds. Since its establishment in 1981, not one breed has become extinct in that country. Moreover, this signal achievement did not use a single penny of public money.

Admittedly, building similar working partnerships among stockraisers, breeder associations, and scientists in the South will not be easy. It will require innovative thinking, a participatory approach, public education, new laws and more. But unless this task is engaged, all the UN conventions, FAO databases and academic or on-station research in the world will not save the relatively more vast livestock biodiversity of the South for food security in the future.

References

Alstrom, S. 1999. The social dimensions of cattle castration in Western Rajasthan. In *Desert, Drought and Development: Studies in resource management and stability.* (eds) R. Hooja and R. Joshi. Jaipur and New Delhi, India: Rawat Publications. 316–324.

FAO. n.d. *The Global Programme for the Management of Farm Animal Genetic Resources: A call for action.* Rome, Italy: FAO.

Hussain, T., P. Bibi and P. Kaushal. 1999. We are all part of the same 'Kudrat' – community forest management in Rajaji National Park. *Forests, Trees and People Newsletter* 38: 35–38.

ITDG. 1996. *Livestock Keepers Safeguarding Domestic Animal Diversity Through their Animal Husbandry.* Rugby, UK: Intermediate Technology Development Group Food Security Programme.

Köhler-Rollefson, I. 1993. About camel breeds: A re-evaluation of current classification systems. *Journal of Animal Breeding and Genetics* 110: 66–73.

—— 1997. Indigenous practices of animal genetic resource management and the relevance for the conservation of domestic animal diversity in developing countries. *Journal of Animal Breeding and Genetics* 114: 231–238.

—— 2000. *Management of Animal Genetic Diversity at Community Level.* GTZ Agrobiodiversity Programme, Eschborn, Germany.

Köhler-Rollefson, I., and J. Bräunig. 1999. LPP's initiative for participatory conservation of indigenous livestock breeds: A concept note. Paper presented at the Deutscher Tropentag, Humboldt University, Berlin, Germany.

Köhler-Rollefson, I. and H.S. Rathore. 1996. The Malvi camel: A newly discovered breed from India. *Animal Genetic Resource Conservation* 18: 31–42.

Martin, M., E. Mathias and C.M. McCorkle. 2001. *Ethnoveterinary Medicine: An annotated bibliography of community animal healthcare.* London: Intermediate Technology Publications.

McCorkle, C.M. 1983. *Meat and Potatoes: Animal management and the agropastoral dialectic in an indigenous Andean community, with implications for development.* Doctoral dissertation. Stanford, CA: Department of Anthropology, Stanford University.

—— 1999. Africans manage livestock diversity. *Compas Newsletter for Endogenous Development* 2: 14–15.

Pereira, F.A. 1998. *Use of World Wide Genetics for Local Needs.* Proceedings of the 6th World Congress on Genetics Applied to Livestock Production. Armidale, Australia. 155–160.

Scherf, B. 1996. *World Watch List for Domestic Animal Diversity* (2nd edition). Rome, Italy: FAO.

Tantia, M.S., P.K. Vij and A.E. Nivsarkar. 1998. *Sahiwal in India.* Paper presented at the Fourth Global Conference on Conservation of Domestic Animal Genetic Resources, 17–21 August. Kathmandu, Nepal.

Timon, V.M. 1993. Strategies for sustainable animal agriculture – an FAO perspective. In *Strategies for Sustainable Animal Agriculture in Developing Countries: FAO animal production and health.* Paper No. 107. (ed.) S. Mack. Rome, Italy: FAO. 7–22.

Vásquez, G.R. 1999. Culture and biodiversity in the Andes. *Forest, Trees and People Newsletter* 34: 39–45.

Chapter 10

Sandy-clay or clayey-sand?

Mapping indigenous and scientific soil knowledge on the Bangladesh floodplains

Paul Sillitoe, Julian Barr and Mahbub Alam

Indigenous knowledge research aims to facilitate the targeting of development resources more effectively on the poor. The compatibility of local ideas with scientific ones is a central issue in this work. One objective is to facilitate communication between scientists and local people, on the assumption, fundamental to development interventions, that science may have something to offer them in tackling their problems. Furthermore, it is possible that if scientific and indigenous knowledge are comparable, and if scientists are able to access local knowledge, this might save on expensive scientific research – on the grounds that sharing what the local people already know may reduce the need to conduct research into some topics – and also facilitate empowerment of the poor – because if their knowledge features prominently in any development initiative this will give them a meaningful role in its planning and implementation.

This chapter's aims are both intellectual and practical. It seeks to compare local Bengali farmers' soil classification with that of soil scientists, to explore parallels and differences. It builds on research undertaken on the Bangladesh floodplain to explore methods for improving natural resources research by combining scientific study of natural resources with farmers' and fishers' local knowledge of resources (Alam 2001; Ghosh 2002; Sillitoe 2000). It takes as its premises that (i) farmers' knowledge of the soils in their fields is the most locally relevant understanding of those soils, and (ii) there are potential efficiency gains over expensive land and soil surveys in collecting and using local soil knowledge. The chapter correlates the mapping of local soil names with a scientific soil survey. It seeks not only to evaluate understandings of soil distribution but also to assess the extent to which a local population's knowledge of its soils might substitute for, or complement, an expensive scientific soil survey. This reverses the usual dialogue in development, by emphasizing local people informing scientists, assessing the extent to which they might communicate intelligence about their soils, so reducing the need to undertake costly pedological survey work; in addition to facilitating the communication of locally perceived problems.

The aims are utilitarian, to improve the relevance of, and reduce the costs of, scientific soil surveys. Scientific land resource and soil surveys are expensive. The era of large area surveys in developing countries is past. Today, with limited budgets, local level small-scale (<1:50,000) reconnaissance surveys are the norm (Tabor, 1992).[1] The incorporation of indigenous knowledge into soil and land resources survey can improve relevance and level of detail (Tabor 1992), and potentially reduce costs. In East Africa, Haburaema and Steiner (1997) propose that the systematic use of farmers' knowledge could yield 'rapid and cheap appraisal of soils of individual fields'. A scientific survey may prove poor value for money if it fails to take account of local knowledge, lacking relevance to local people (Tabor and Hutchinson 1994; Sillitoe 1998) and lead to poor land use planning and management decisions (Kundiri *et al.* 1997).

It is intuitive that those making a living from the land are well placed to describe that land and its soils, albeit using their own location specific terminology. The recent burgeoning of ethno-science literature on soil testifies to it (Sillitoe 1996; Talawar and Rhoades 1998; Winkler Prins 1999). We have yet to demonstrate how we can simultaneously utilize this knowledge and scientific soil knowledge (Payton *et al.*, 2003). In order to promote the use of local soil knowledge, we have to make it accessible to researchers, policy makers and development professionals. The key challenges in providing information to support sustainable development are to ensure acquisition costs are low and utility is high (Tabor and Hutchinson 1994).

Geographic information systems

Besides trying to distil such knowledge in reports, we have the option today of new information and communications technology, which we can integrate with computerized knowledge systems (Gonzales 1995). One such system with the potential to combine indigenous and scientific soil knowledge is the geographic information system (GIS). It comprises a series of digital maps that store various data about a location in layers that one can combine, interrogate and analyse. Anthropologists have made some use of GIS (Aldenderfer and Maschner 1996), as have others in research on farmers' indigenous knowledge (e.g. Lawas and Luning 1996). The propriety of investigating indigenous knowledge using geographic information systems is contested.

Supporters argue that indigenous knowledge can be 'quantified and systematically organised by means of a GIS', that it makes access easier, promotes data sharing and analysis (ibid). It makes it easier to store, disseminate and replicate the intelligence (Harmsworth 1999; Gonzales, 1995). Others point out that GIS could help establish security of land tenure and resolve boundary disputes (Mohamed and Ventura 2000). The technology has illuminated social construction of environmental understanding when used to document and compare contrasting perceptions of natural resources and access (Weiner *et al.* 1995).

Detractors raise concerns that GIS's demand for certain kinds of data may impose a narrow orthodoxy, and for local people an alien one, on the design of development projects (NRI 1999). It is an expensive technology, hungry for high quality data, and needs trained users (Campbell 2002). It privileges Western cultural perceptions of space, imposing a positivist perspective on the world, such that the cartographer's view of the landscape dominates (Pickles 1995). Furthermore the technology places the GIS user/owner in a powerful technical gatekeeper position, leading to concerns about unethical use (Dunn *et al.* 1997).

While we recognize the dangers of reifying indigenous knowledge in databases (Barr and Sillitoe 2000: 190) and manipulating it remote from its owners, we purposively treat indigenous knowledge in this chapter as a form of data that one can manage using GIS, in a manner similar to spatial data collected in a formal survey. We take this positivist stance to explore whether GIS can utilize and add value to relatively low cost ethnographic information about soils, and to assess the potential for utilizing local soil knowledge to produce soil maps, as scientists use survey data. We wish to see if this is technically feasible, and gauge drawbacks to its widespread use, as a potentially inexpensive way to conduct soil surveys that meaningfully involves local people and draws on their expertise. If we are to achieve these objectives, we need a method that allows us to extrapolate from individual farmer's expert knowledge of their own fields to the broader coverage found in soil maps. This chapter describes and appraises a method for deriving such an indigenous soil map from point-specific farmers interviews.

Field site characteristics

The research was conducted at a site on the Jamuna floodplain in the Tangail district of Bangladesh, approximately 16 kilometres west of the Jamuna River. It is located between two river systems – to the west, the Dhaleswari, a major distributary of the Jamuna, and to the east, the Bangshi, which drains the slightly uplifted Madhupur Tract (EGIS and Delft Hydraulics 1997). It comprises a shallow saucer-like floodplain depression, known as a *beel*, which includes a small perennial waterbody of 44.5 hectares at its centre, surrounded by arable land that is seasonally flooded as the waterbody expands to cover 394 hectares during the monsoon. Settlements are located on higher land around the margins of the depression. The local administrative unit (*mouza*) is divided between two Agro-Ecological Zones (AEZ): AEZ 8, the Young Brahmaputra and Jamuna floodplains, and AEZ 9, the Old Brahmaputra floodplain (FAO and UNDP 1988). The study site lies just to the east of the boundary between the AEZs, and is in AEZ 9 (Office of Field Services 1993). The soils of the area are predominantly non-calcareous grey floodplain soils developed from alluvium.

The soil surveys

Two independent surveys of the soils of the area were undertaken: a local soil knowledge survey featuring open ended questions in farmers' fields and formal soil survey with physical assessment using auger and profile pits. The entire study area covered 871 hectares, with 4,730 paddies or cadastral plots across it. Neither of the two surveys, of formal soil classes or local soil knowledge, covered the whole area, nor shared identical footprints. When un-surveyed administrative units (*mouza*), and the area occupied by homesteads and ponds are removed, the remaining common area covered by both surveys was 2,708 cadastral plots, or 503 hectares, on which this investigation focuses.

Indigenous soil survey

There is a range of methodologies for collecting indigenous soil knowledge. Common ones include interviewing individual farmers or holding group discussions. One cannot readily use such information to produce a map, to demarcate spatially farmers' knowledge of soil types. Farmer-drawn sketch maps, or PRA maps (Furbee 1989; Birmingham 1998) are commonly used currently in development contexts to collect such data. Participants may draw maps free hand, or plot their knowledge onto formal base maps of differing detail (from basic maps showing features such as rivers and roads through to transparent film over aerial photographs). Alternatively, researchers may interview farmers at known locations, such as in fields, geo-referencing the interview place using a hand-held global positioning system (GPS) device and plotting it on a topographic map (Payton *et al.*, 2003). Several sketch maps may be combined to give a map of a useable extent or responses from geo-referenced points be interpolated. There are dangers of distorting farmers' knowledge, particularly in identifying boundaries between indigenous soil units.

Farmers are very knowledgeable about the land that they cultivate, that is their fields or rice paddies, which is based on accumulated experience. Regarding soils located in fields/paddies that they do not cultivate, their information becomes less reliable as it depends more on deduction and less on experience. We can avoid this by only asking farmers about the soils they cultivate in their own fields or paddies, although this involves more work interviewing many farmers to obtain enough information on sufficient plots to create a mosaic map. There is also the problem of standardizing responses from many interviewees. We used this method in this study.

In Bangladesh, as in much of India, the colonial authorities produced remarkably detailed (1:3,960) cadastral maps of land holdings for use in land taxation, settlement of boundary disputes and regularizing land sale in association with deed documents.[2] Farmers today are familiar with these maps, and many can readily locate their rice paddies on them. The boundaries on these maps are permanent: they exist in fields, usually comprising low earth ridges (called *ails* or *bundhs*), constructed to maintain irrigation water in rice paddies. We can accurately locate farmers' information, when asked about their soils, on these maps, matching it with precisely demarcated areas where they have a fund of practical knowledge tilling the land.

Two local young male research assistants, directed by a Bengali post-graduate anthropologist resident in their community, and overseen by visiting supervisor assisted by other project staff, conducted the indigenous knowledge survey in 1998 and 1999.[3] We approached farmers working in their paddies and put the same question to them: 'What do you call the soil here?' (*'Ekhankar matik-eye apnara ki mati bolen?'*). We sought the name of the soil in the paddy where they were at the time – plots averaging 0.19 hectare (ranging from 0.004 to 13.3 hectare). We asked a random sample of farmers to name the soil type in their paddy plots, analogous to a scientific free survey. The survey was random in the sense that the sample depended on farmers who happened to be working in their fields at the time. We did not question local key informants on soils, as people assert that there are no soil specialists, that such knowledge is more or less equally distributed between all farmers.[4] We thought that we should obtain a better representation of soil knowledge by sampling widely, reflecting its cultural distribution, assuming that those farming particular plots would know the soil there better than anyone else.[5] We noted the results of these short interviews spatially, locating plots on the cadastral map and tabulating responses accordingly. We subsequently used these data to create a GIS indigenous soil knowledge map layer, plotting the soil types in the many isolated survey paddies. In addition, we conducted in-depth open-ended interviews with farmers over the two years of fieldwork, to obtain an understanding of local soil concepts, and the content of the names.[6]

While this method is sensitive to the cultural position, it poses certain problems. It yields a body of non-standard responses. Whereas one or two informants will soon fall into the habit of using a standard set of terms, when one asks many different people the same question one is likely to obtain a considerable range of responses. It was evident in the responses that different people may interpret the question 'What do you call the soil here?' in different ways. While some replied with terms which we recognized as *mati* (soil) names, which we were after, others initially replied with neighbourhood names, such as 'this is the soil in the Red Lentil Neighbourhood' or topographical terms, such as 'this is the soil on a homestead ridge'. We followed up such responses with another question stressing that we were after the name of the soil. Another related problem was that people sometimes used different terms for what they agreed was the same soil. This is usual in ethno-scientific work and as our enquiries progressed we sorted out these synonyms. More problematic than synonym variation, we found that when we asked more than one person to identify a soil they sometimes disagreed over the class to which it belonged (we were unable to assess as planned the extent of such disagreement over naming of soils because of illness).[7] Again, while this is a problem for a survey of the sort attempted here, it reflects cultural reality, for people frequently disagree over the identification of natural phenomena.

Formal soil survey

When we were planning the study, the soil scientists expressed concern about the fluidity of the indigenous soil survey methodology. They proposed that one or two local people should accompany the surveyors to give the Bengali name for soils inspected. After some discussion they conceded that such an approach would fail to capture variable local understanding. Furthermore we pointed out that the dominant soil science model would likely influence indigenous responses, as the one or two local informants shadowing the soil survey became familiar with the surveyors' methods and keyed-in to their expectations, employing a standard set of mirror terms thinking that this was what we required of them, i.e. to reflect the scientists' ideas. The result would be an artificial correlation between scientific and local soil conceptions.

The soil survey comprised two parts, systematically to assess and map the soil resources of the Charan *Beel* catchment (McGlynn and Payton 1998a; Payton *et al.*, 2003). First, a team of Bangladeshi soil scientist colleagues carried out a reconnaissance survey in 1997, resulting in an initial map legend using the soil series associations described in the 1:125,000 Reconnaissance Soil Survey of Tangail District (Brinkman 1967). Second, a joint team of European and Bangladeshi surveyors carried out a detailed survey in 1998. They conducted the detailed survey at a scale of 1:3,600, using the local 16 inch:1 mile cadastral map of the area as a base map. They employed standard free survey procedures (Dent and Young 1981), using an Edelmann auger, to sample extensively and establish the distribution of soils, and their relationships to topography, updating the map legend as they proceeded and digging reference profile pits at key locations as necessary to identify and describe the different soil types (McGlynn and Payton 1998a).

The methodological and conceptual differences between the two surveys, and the knowledge traditions upon which they depend, compromise at the outset in some measure our aim of comparing scientific soil survey with local knowledge of soil distribution. In some significant regards they are not comparable. We pursued the GIS based analysis in full knowledge of the epistemological differences between the two bodies of soil knowledge, to assess the extent of comparability/incomparability between them, subjecting both to investigation using the same formal scientific tool of analysis, on the grounds that this might be a way of more effectively incorporating indigenous soil knowledge into development interventions to the benefit of both local people, by empowering them, and funding agencies, by pioneering more cost effective survey approaches.

The soils

The most striking finding of the surveys is the absence of any readily seen pattern in the distribution of soils (Figures 10.1 and 10.2). The scientific map shows no catenary distribution of soil series, as we might expect in a sloping landscape, with instead patches of different soils. And there is no pattern evident in the local map coverage, soils described as sticky and sandy occurring in adjacent plots.

The generic local term for soil is *mati*. The word may derive from *ma* (mother), Bengalis today refer to the soil as mother. Farmers distinguish between and name several soils, and from other research, there appears to be considerable regional variation, both with respect to dialect regarding widely distributed soils and geography regarding the localized occurrence of some soils (Sillitoe 2000; Ghosh 2002; McGlynn and Payton 1998b). The criteria used to

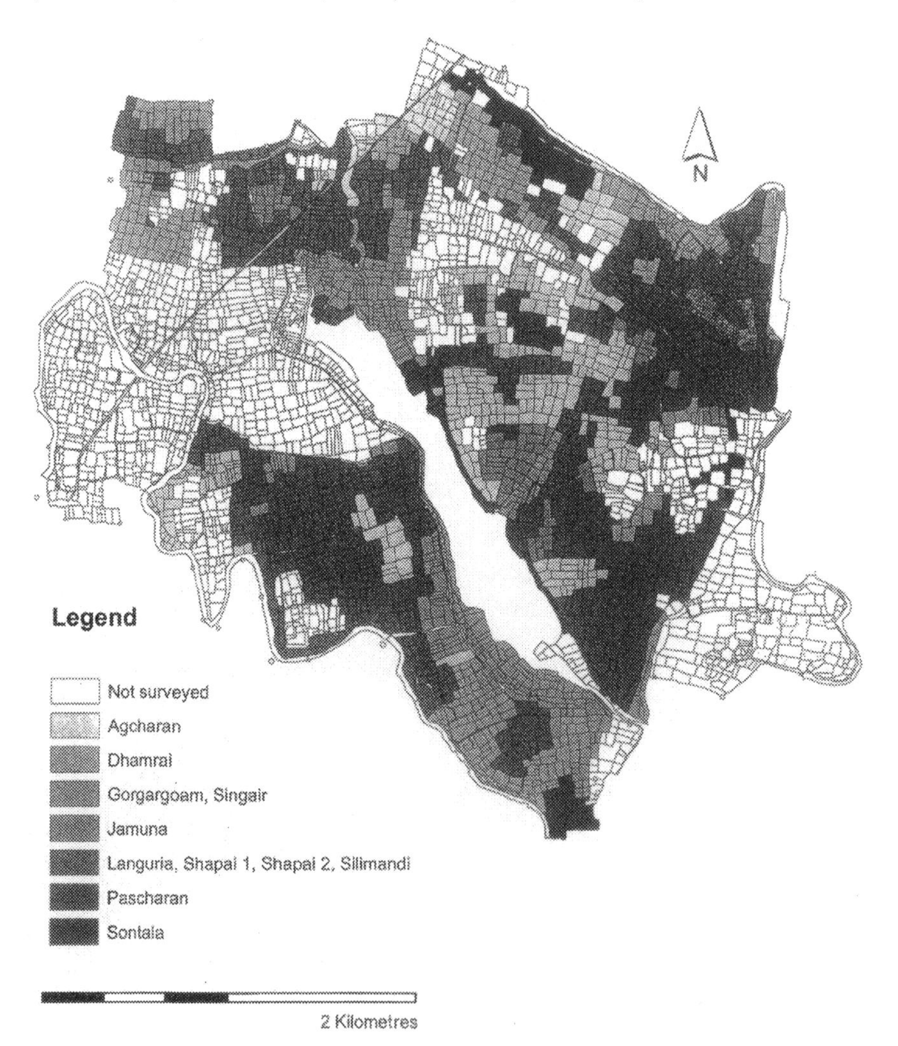

Figure 10.1 Scientific soil map of Charan *Beel*. A full colour version of this figure can be found at http://lucy.kent.ac.uk/lkder/Sillitoe

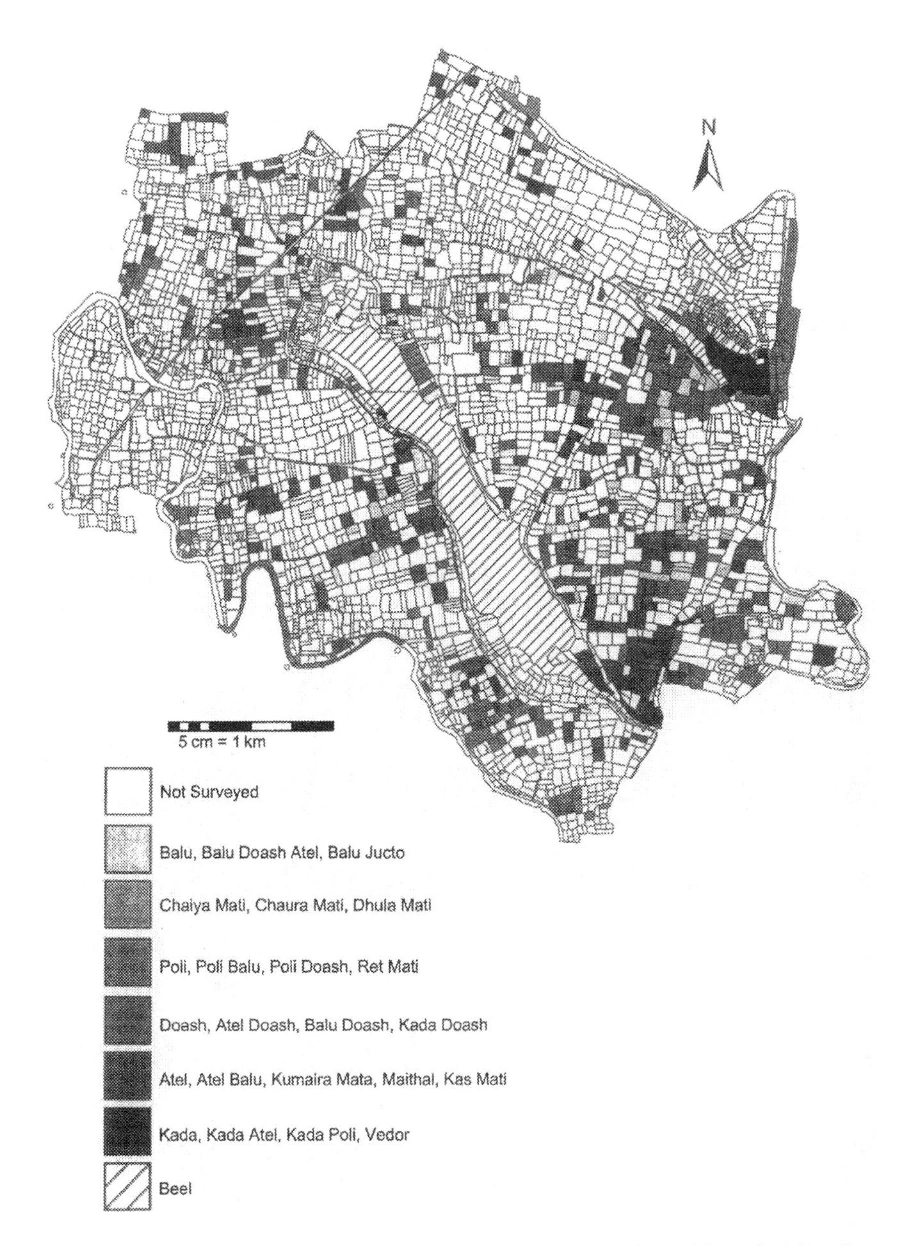

Figure 10.2 Map of indigenous classification of soils in single paddies. A full colour version of this figure can be found at http://lucy.kent.ac.uk/Ikder/Sillitoe

distinguish soils vary. People see a range of cues, although they frequently refer to texture and colour, in addition to physical location. There appears to be some association in their minds between landscape position and soil type. This was evident in the interviewing process, when farmers often initially responded with the landscape position (e.g. high land) rather than the soil name.

Table 10.1 gives the key properties of the 21 indigenous soil types, ascertained through farmer interview. The two most extensive local soil types in the study area are *atel* and *balu*, followed by *balu doash*. These occupy 32.6 per cent, 31.9 per cent and 12.0 per cent of the GIS map grid squares respectively. There is an expectation that *atel* clayey soils will dominate closer to the perennial *beel* and that sandier *balu* soils will occur on levees and moderately flooded land.

The soil class *atel* is very sticky and plastic when wet (the name *atel* derives from the word *atha* for sticky) and difficult to remove from tools. It has slow permeability and farmers consider it a good rice soil. When dry it becomes a hard, compact and strong soil that is difficult to plough. It cracks and produces hard clods (*dhel*) in the dry season, which farmers sometimes have to break with a wooden mallet before the next cultivation.

The soil class *balu* or *bailla* is a coarse textured soil. It remains loose (*jhor-jhore*) and easy to plough even in dry conditions. But it is droughty and lacks moisture (*rasnai*). It does not retain water well, and consequently is not considered a good rice soil.

It is common practice to combine two soil terms to describe soils that combine characteristics of both. Farmers recognize a class of mixed soils called *doash*, which literally translates as *do* (two) and *ash* (fibres). It is a loamy soil in which sand and clay are present in more or less equal proportions. If the proportion of sand is higher people speak of *balu doash*, these sandy loam soils are always easy to cultivate. A soil with a small *balu* sand content is called *balu jucto*, these clayey loams can be more difficult to cultivate in dry conditions. These are textural classes. The Charan farmers also distinguish a pale form of *doash*, which they call *chaiya chaiya*. They identify it by its pale ashy colour, likening it to ash – the name derives from *chai* the Bengali word for ash, and like ash, this soil blows in the wind.

When farmers describe a paddy as having *atel* and *balu* or *balu* and *atel* soil, this indicates that both soil types are present in their pure form, each occupying a separate part of the paddy; the first named in the larger proportion. Where a soil features both *bele* and *atel* unmixed, that is comprising two separate *vaj* (layers) one on top of the other, people refer to it as *kas mati*.

Other clayey soils in the area include *kumaira mati*, a black and slippery soil (people easily fall over walking on it) that is used as a raw material by the *kumar* potter caste. And *vedor*, a very soft, sinking soil that occurs at the *beel* margins and underwater. It is reported to smell of rotting vegetation, and people say that it is a particularly fertile soil because of its high organic matter content.

The silt deposited by floodwater, which may be a gleyed blue colour in the subsoil, is called *poli mati*. It may comprise a thin layer (four fingers or less

Table 10.1 Indigenous soil categories

Indigenous soil name	*Key characteristics*	*Comments derived from farmer interviews*
Atel	Fine textured, sticky soil – sticks to feet	Fertile (*jore beshi*); high yielding; good for wheat, mustard, *boro* rice, all crops; good water holding properties (up to eight days between irrigations); full of *ras* (literally sap, i.e. moisture); soft when moist but hard when dry. Soils become more sticky (*atailla*) with cultivation. One of the best soils.
Atel + balu	Clayey and sandy	Farmers describe this as the soil in their paddies where both *atel* and *balu* present, and *atel* is dominant in extent.
Atel doash	Clay loam	*Atel* mixed with any other soil type, but *atel* dominates. Very little sand content; good for IRRI *boro*, mustard and *aman*; hard to cultivate; reasonable water holding capacity; farmers apply 2 kg fertilizer/decimal; higher yielding and better water holding than *bele doash*.
Baiala, Bali, Bele	Coarse textured	Synonymous with *balu*.
Balu	Coarse textured – poor water holding	Less fertile (*jore kam*); low yielding due to little *ras* (*rasnei*); the *ras* lasts only a month after monsoon flooding; the worst soil; poor crop growth even with heavy use of irrigation and fertilizer; can be blown by wind (*atar jomi* – like wheat flour); good for mustard but not IRRI *boro* rice; crops are harmed if irrigation is infrequent, irrigation required every two to three days; high fertilizer requirement.
Balu + atel	Sandy and clayey	Farmers describe this as the soil in their paddies where both *balu* and *atel* present, and *balu* is dominant in extent.
Balu jucto	Sandy	Any type of soil in which one can feel sand.
Chaiya chaiya mati	Literally 'ash', a fine, blowing, pale soil	
Chaura mati	Sand in *char* – fine and slightly wet, blows, but not like *chaiya*, large grains	
Dhula mati	Very fine textured soil, literally 'dust'; not sandy, no sandy texture	

continued

Table 10.1 continued

Indigenous soil name	*Key characteristics*	*Comments derived from farmer interviews*
Doash	Medium textured, when two soil types are mixed together (from *do* the root word indicating 'two')	The best of all soils, especially for mustard and IRRI *boro* rice, suitable for all crops; very fertile (*jore beshi*); provides a good yield; a mixture of *atel* and *balu*; contains a little sand (*balu*), contains *atel* and *ret*, is full of *ret*; requires irrigation every three days; keeps its *ras* (moisture) for a long time; the preferred soil for jute.
Kada	Very sticky clay	One of the best soils because it does not require irrigation or fertilizer; good for local *boro* rice; located within the *beel* and always underwater, soft, but very hard if it does dry out; dark, black. Similar or associated with *vedor* soil.
Kada + atel	Very sticky clay and less sticky clay	Both soils occur in the paddy; *kada* predominates.
Kas mati	Type of *atel*, but with separate horizons of *atel* and *balu*	Maybe synonymous with *domasailla*; also described as a sub-type of *atel*; holds water for up to four days.
Kumaira mati	Potters' clay (used for making earthenware pots) – a sub-type of *atel*	Very hard to plough; the hardest type of soil; pure *atel* (*pakka athailla*); soft and 'greasy' when wet; like soap when laddering it; quite fertile.
Maithal	Fine textured – a sub-type of *atel*	Has the characteristic of *domasailla*/*doash* – made up of mainly *atel* with a little *balu*; highland (*vita*) soil – occurs around the homestead.
Poli doash	Silty mix	
Poli mati	Silt	
Ret mati	A reddish silt deposited in flood water, not sticky – used as potters' slip	Best for *boro* rice; good for mustard; like *atel*, but stays wet, the *ras* stays for a long time; one farmer said originally the whole of Charan was *ret mati*, 'but the river brought *poli* and the *balu*.' Deposition of *ret* increases fertility; more is deposited closer to the *beel*.
Vedor	A sinking/liquid type of clay	Forms when *atel* is under water for long periods; very soft, non-weight bearing, people can sink past their waists in it – has '*dab*' (literally propensity for things to sink in); may be inundated all year.

deep). Many farmers comment that it is deposited in far lower quantities today than before flood control structures were constructed on the floodplain. At Charan *beel* people also distinguish a reddish silt called *ret mati*, which they say the floodwater brings and deposits from the adjacent uplifted Modhupur Tract. Potters use it as slip to colour pottery red.

Table 10.2 gives the key properties of the 11 scientific soil types, ascertained through soil survey. According to the scientific soil classification system of the FAO-ISRIC-ISS (1998), all of the soils in the Charan study area, with the exception of one, are *gleysols*. The exception is the soil series *pascharan,* which falls within the *hapli gleyic arenosol* group. It is the most extensive soil in the Charan *beel* area, occupying 21.8 per cent of the GIS map grid squares. It is very coarsely textured and is found mainly along the banks of the *beel* on low levees and in several patches elsewhere. The *agcharan* soil series is also coarsely textured, but differs from *pascharan* in having a finer textured surface horizon. Both these soils are very permeable and their poor water holding capacity renders them problematic for paddy rice production (Barr and Gowing 1998).

The other soil series of significant extent are *jamuna*, *dhamrai*, *gogargoan*, *languria* and *shapai 1*, which occupy 16.1 per cent, 15.2 per cent, 12.4 per cent, 11.9 per cent and 8.7 per cent of the GIS map grid squares respectively. The *jamuna* and *dhamrai* series soils occur on levees, the former being a sandy loam over sand, and the latter a strongly structured silty or clayey loam. The *gorargoan* series soils occur in basin depressions and are nearly permanently waterlogged clays over blue-green silt loam. In general, grey alluvial soils with sandy clay to sandy clay loam surface texture, over sandy clay and coarse blocky to prismatic structured subsoil, dominate the mid- and lower slopes of the shallow Charan *beel* floodplain depression (McGlynn and Payton 1998a). Many of the soils in the study area have unstructured sand at or below 100 cm, which allows for rapid movement of ground water.

The GIS analysis

We used a GIS to store the formal and local soil maps resulting from the surveys. We created a GIS base layer demarcating the boundaries of all the individual paddy plots from the 1:3,960 cadastral map. We made a further two layers, at the same scale. One from the scientific soil survey of the study area, the other from the indigenous soil data from 413 largely isolated plots (15.3 per cent of paddies). We compared the two soil layers for spatial correspondence between the two classification systems and to investigate the validity of interpolating local soil knowledge from point data.

The GIS comparison involved intersecting the two survey areas, to derive the common area covered by both soil layers. We used this area as an analysis mask. We converted the soil unit polygons within the mask area of the scientific vector soil map to a raster grid[8] using a 5 m resolution. We also changed the individual and frequently isolated plots on the indigenous vector soil map to a raster grid

Table 10.2 Scientific soil categories

Soil Series	*FAO-ISRIC-ISSS (1998)*	*Landform position*	*Topsoil characteristics*	*Subsoil characteristics*	*Soil drainage/ water regime*
Agcharan	Hapli-endoarenic gleysol	Low floodplain ridge/levee	Greyish brown weak subangular blocky mottled sandy loam.	Light grey massive or single grain fine loamy sand within 50 turning to fine sand within 75 cm.	Poorly drained very permeable seasonally water logged by high groundwater.
Dhamrai	*Eutri-fluvic gleysol*	Low floodplain river levee	Olive grey mottled weak coarse blocky silt loam to silty clay loam.	Dark grey mottled sandy clay loam with moderate strong structure.	Poorly drained, permeable seasonally waterlogged by high groundwater.
Gorargoan	*Hapli-humic gleysol*	Floors of floodplain basin depressions	Very dark grey mottled clay loam to sandy clay loam with massive to weakly coarse blocky structure.	Dark grey to black finely mottled clay to clay loam with moderate blocky structure over a blue green massive silt loam.	Very poorly drained subsoil permanently water logged by groundwater.
Jamuna	Hapli-dystric gleysol	Low floodplain ridge/levee	Greyish brown weak to moderately strong blocky sandy loam to loam over a strong blocky greyish brown mottled sandy clay loam.	Grey weak to loose blocky sandy loam turning to loose fine olive sand within 100 cm.	Poorly drained Permeable seasonally waterlogged by high groundwater.
Languria	*Hapli-eutric gleysol*	Mid-slope of floodplain	Grey to dark grey mottled sandy clay to sandy clay loam with moderate to strong blocky structure.	Grey to olive sandy clay loam to sandy clay with moderate blocky structure over massive sand.	Poorly to very poorly drained, permeable, seasonally waterlogged by high groundwater.

Pachcharan	*Hapli-gleyic arenosol*	Low floodplain ridge marginal to Charan *Beel*	Light grey loose loamy sand.	Massive loamy sand stratified with grey sandy loam.	Poorly drained very permeable seasonally waterlogged by high groundwater.
Shapai 1	*Hapli-eutric gleysol*	Mid-slopes of floodplain ridge	Greyish brown mottled sandy clay to clay loam with moderate to strong blocky structure. Very porous.	Grey mottled sandy loam with coarse blocky to prismatic structure.	Poorly drained, moderately permeable, seasonally waterlogged by high groundwater.
Shapai 2 (sandy subsoil phase)	*Hapli-eutric gleysol*	Mid-slopes of floodplain ridge	Very shallow greyish brown mottled sandy clay loam or clay loam with moderate to strong blocky structure. Very porous.	Grey mottled sandy loam between 25 and 50 cm with moderate blocky structure over massive fine to medium loamy sand within 100 cm.	Poorly drained, permeable seasonally waterlogged by high groundwater.
Silmandi	*Hapli-eutric gleysol*	Mid- and lower-slopes of the floodplain.	Dark greyish brown clay loam to clay with strong brown mottles and strong coarse blocky structure.	Brown mottled silt loam to silty clay loam with weak coarse blocky structure over loamy sand at depths greater than 150 cm.	Poorly drained moderately permeable, seasonally waterlogged by high groundwater.
Singair	*Hapli-humic gleysol*	Fringes of floodplain basin depressions	Grey mottled clay to clay loam with massive to weak coarse blocky structure.	Grey to dark grey mottled clay to sandy clay loam with strong blocky structure over loamy sand or sand at approximately 100 cm.	Very poorly drained subsoil water logged for long periods by high groundwater.
Sonatala	*Hapli-fluvic gleysol*	Low floodplain river levee	Light grey mottled weak very coarse blocky silt loam.	Grey to dark grey mottled silt loam with strong platey structure. Stratification evident.	Poorly drained, permeable, seasonally waterlogged by high groundwater.

Source: After McGlynn and Payton (1998a).

format. We then used a proximity analysis function in Arcview GIS to interpolate the grid squares with indigenous data, to achieve full map coverage, again at 5 m resolution. The interpolation was based upon the qualitative attribute of local soil name (Figure 10.3). The resulting two GIS layers comprised a scientific and a local grid map, each of 655 rows and 720 columns (i.e. a grid of 471,600 cells covered the area of both soil classifications).

We compared the derived indigenous map (Figure 10.3) with the scientific

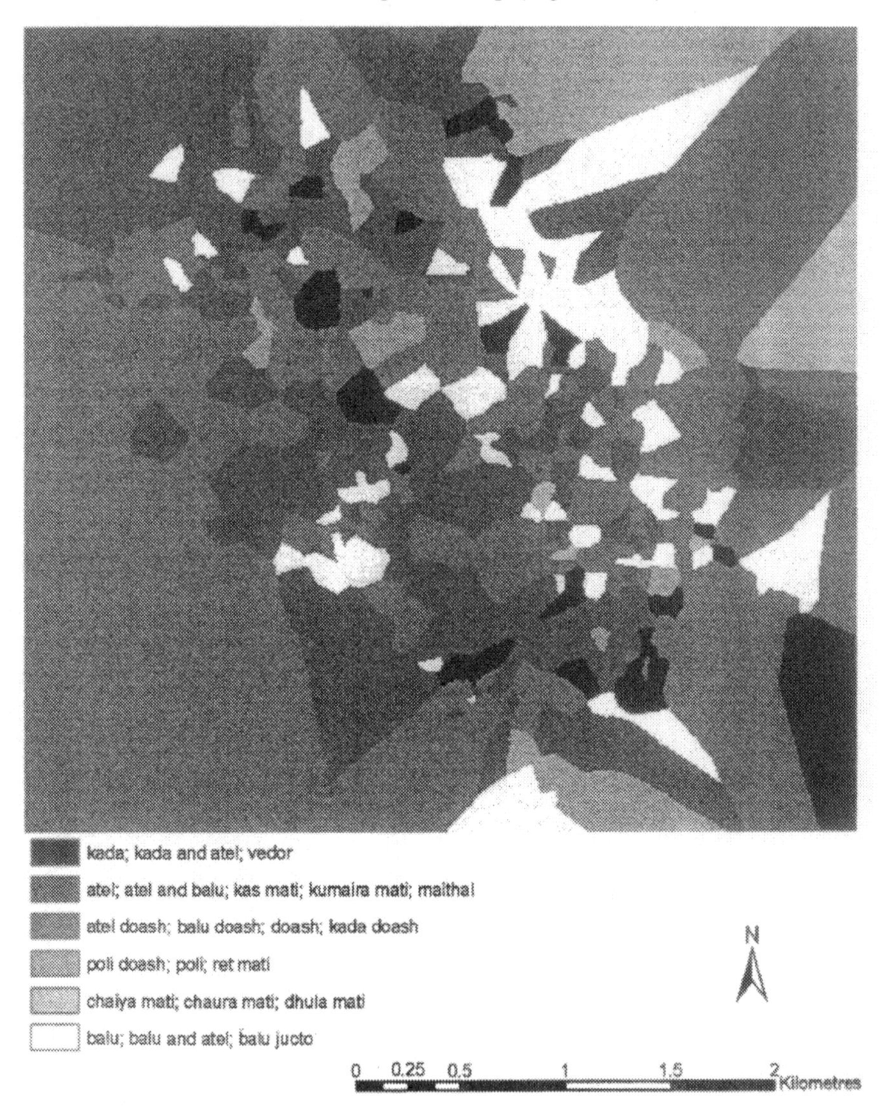

Figure 10.3 Interpolated full coverage indigenous knowledge soil map. A full colour version of this figure can be found at http://lucy.kent.ac.uk/Ikder/Sillitoe

one of the same area at the same scale using correspondence analysis. We combined the plot attributes from the two grid layers into a three-column table (grid cell identification, scientific survey class, indigenous knowledge class) with 471,600 rows (each row representing a grid cell). We used GRASS GIS software to calculate a coincidence table from these data, using the *r.coin* function. The resulting 11 × 21 cell coincidence table gave us the number of observed coincidences in the 5 m × 5 m GIS grid of all the possible combinations of the 11 scientific and 21 local soil classes (Table 10.3).

We can test for independence in 2 × 2 coincidence tables using the chi-square test, whereas to determine departures from independence in larger tables with more than one degree of freedom, such as in this 11 × 21 table, demands more detailed analyses (Everitt 1992: 37). We calculated a table of adjusted residuals for the 11 × 21 coincidence table to identify the cells where the two soil classifications correspond with significant overall chi-square scores (ibid: 47–48). The coincidence table significantly disproves the null hypothesis of independence between the two soil classifications ($\chi^2 = 127956$, 200 DF, $p < 0.001$). Table 10.4 of adjusted residuals shows 15 local-scientific combinations with high residual scores.

We might expect the two soil classifications to show considerable coincidence as both farmers and soil surveyors are describing the same substance. There are a few good statistical correlations between scientific and indigenous soil classes, where the soil descriptions show a physical basis for these relationships. The coincident pair with the highest adjusted chi-square residual value – *kas mati* and *shapai 2* (pair 1, Table 10.4) – show good agreement over soil type. Farmers describe *kas mati* as a fine textured soil comprising separate layers of *atel* (clay) and *balu* (sand). The soil scientists classify *shapai 2* as a *gleysol* with a clay loam surface horizon and sandy loam subsoil. This coincident pair covers only 0.13 per cent of the survey area (occurring in only 609 grid squares). The farmer-surveyor descriptions of some other pairs (6, 7 and 10 to 15, Table 10.4) also show some similarities, although not with the level of agreement seen in pair 1. Both local and scientific classifications allude to coarser texture regarding pair 6, both identify a blocky clay for pair 12, and both classifications describe clay-sand mixtures for pair 15. Overall, the soil descriptions that show some correspondence tend to have coarser textures (pairs 1, 6, 7, 11, 13, 15), although we cannot make much of this observation, as none of the high residual pairs include *pascharan*, the sandiest of the surveyors' soil series.

Examination of other coincident pairs reveals that they do not predictably relate together soils that according to the two classification systems have similar features. Other coincident pairs with high adjusted residual values (pairs 2 to 5, Table 10.4), are all mismatched. The first pair – *ret mati* and *agcharan* – farmers describe as reddish silt deposited in floodwater, and surveyors as a greyish brown sandy loam over loamy sand. The next pair – *vedor* and *sontala* – they describe as a soft clay with high organic content, and a mottled grey silt loam, respectively. The next two – *atel doash* and *agcharan* – are a clay loam

Table 10.3 Coincidence table. Number of 5 m^2 grid squares in which local/scientific combinations occur

	Agcharan	*Dhamrai*	*Gorgargoam*	*Jamuna*	*Languria*	*Pascharan*	*Shapai 1*	*Shapai 2*	*Silimandi*	*Singair*	*Sonatala*	*TOTAL*
Atel	0	0	186	651	1,859	30	337	0	0	0	12	3,075
Atel + balu	1,887	8,164	6,967	6,717	9,091	18,300	8,817	803	1,978	910	592	64,226
Atel doash	1,435	83	841	621	417	1,210	0	0	469	21	389	5,486
Baiala	271	700	1,664	1,175	2,335	608	1,995	0	339	356	278	9,721
Balu	912	11,432	8,744	17,911	7,110	12,085	1,476	1,665	1,108	2,155	978	65,576
Balu + atel	0	352	1,040	125	665	1,514	993	1,088	350	195	0	6,322
Balu doash	1,401	5,747	2,621	1,879	207	6,935	2,641	1,121	59	1,041	499	24,151
Balu jucto	0	467	0	936	0	1,101	14	26	0	0	0	2,544
C.c. mati	0	39	0	0	0	9	157	0	0	0	0	205
Chaura mati	52	1,514	47	404	764	360	0	0	0	0	158	3,299
Dhula mati	0	438	0	0	0	119	196	19	0	178	0	950
Doash	0	583	0	99	0	239	0	289	0	0	0	1,210
Kada	286	721	739	1	0	229	97	0	0	636	656	3,365
Kada + atel	0	0	0	1,033	98	0	0	0	0	0	0	1,131
Kas mati	0	19	0	0	0	162	0	609	0	14	0	804
Kumaira mati	0	174	183	0	0	86	0	0	0	0	0	443
Maithal	0	0	728	478	937	287	42	0	49	0	522	3,043
Poli doash	0	0	0	0	0	175	199	161	0	0	0	535
Poli mati	0	221	0	0	0	50	0	0	0	0	0	271
Ret mati	1,264	0	587	341	335	360	324	0	0	0	0	3,211
Vedor	0	0	592	0	241	0	298	0	0	0	633	1,764
Total	7,508	30,654	24,939	32,371	24,059	43,859	17,586	5,781	4,352	5,506	4,717	20,1332

Table 10.4 Local-scientific soil combinations with high-adjusted residuals

Pair no.	*Indigenous soil type*	*Scientific soil type*	*Adjusted residual*
1	*Kas mati*	*Shapai 2*	121.9
2	*Ret mati*	*Agcharan*	105.4
3	*Vedor*	*Sontala*	92.0
4	*Atel doash*	*Agcharan*	86.7
5	*Atel*	*Languria*	79.3
6	*Balu*	*Jamuna*	71.7
7	*Balu + Atel*	*Shapai 2*	67.3
8	*Kada*	*Sontala*	65.0
9	*Kada + Atel*	*Jamuna*	63.1
10	*Kada*	*Singair*	56.7
11	*Balu*	*Shapai 1*	−56.2
12	*Kumaira mati*	*Sontala*	53.4
13	*Balu doash*	*Languria*	−49.9
14	*Chaura mati*	*Dhamrai*	45.1
15	*Doash*	*Shapai 2*	43.1

versus a greyish brown sandy loam. And the final pair – *atel* and *langulia* – are a sticky clay, and a mottled dark grey sandy clay. Other pairs with lower adjusted residual scores that show similar mismatches are 8 and 9. The first, farmers describe as a very sticky clay and surveyors as a mottled grey silt loam, and the second a sticky clay versus a sandy loam.

Same soil, different perceptions?

What is the basis for coincidence, or lack of it? A possibility is that the indigenous and scientific knowledge and classification systems do not correspond well. Anthropology has long addressed the commensurability of what representatives from different cultures understand. Some of these investigations, notably in ethno-science and behavioural psychology, have sought to further our understanding of human cognitive processes, seeking to establish the extent to which we understand things and process knowledge in the same way, regardless of culture. One of the tenets of this work is that, assuming that we can agree that tangible things exist out there independently of our minds, human beings should in some physical senses perceive of them the same way, and regardless of their cultures' formulations of this perception there should be some comparability.

Regarding the natural world, this cognitive work has focused largely on discrete phenomena, such as plants and animals. Some claim, for instance, that all humans discriminate at the genera/species level, that this represents a basic point of recognition – for example all of us will see a blackbird as different from a robin or an eagle (Berlin 1992; Lakoff 1987; Rosch 1975). There is some dispute about the comparability of resulting taxonomic arrangements, whether hierarchical arrays are universal (Sillitoe, 2002). While no two members of a single plant or animal species/sub-species are identical, overlapping with one another and nearest neighbours (e.g. sympatric species), they unarguably exist separately. Different soils comprise continua to a far greater extent, grading one into another. They lack discrete natural boundaries that could readily determine their division into classes, such that their classification could possibly be more culturally dependent. Soils are more akin to colours in their continuity, although considerably more complex in composition; cognitive scientists have found cross-cultural variation in divisions of the spectrum (Berlin and Kay 1969; Rosch 1975).

Several researchers have studied the way in which local people classify their soils. The extent of correspondence between their classifications and those of scientists depends upon the comparability of the criteria each uses. Sikana (1993) reports little correlation between local and scientific soil classification systems in Zambia, which he attributes to the different criteria they use. Local knowledge focuses mainly on observable topsoil characteristics that affect management for cultivation, whereas scientific soil knowledge depends on chemical and physical criteria. Haburema and Steiner (1997) report similar lack of correspondence and point out that soil productivity, and hence topsoil properties, interest farmers, while scientists use universal soil classification systems that employ a wide range of diagnostic criteria. Other studies have shown that the criteria farmers use to classify their soils include landscape position, texture, consistency, colour, humidity, hardness, depth and fertility/productiveness (Furbee 1989; Sikana 1993; Gonzales 1995; Sillitoe 1996; Habarurema and Steiner 1997; Kundiri *et al.* 1997; Gobin *et al.* 2000). The most widely cited criteria are soil textural characteristics, which relate also to productivity, soil water retention and workability.

Farmers in many countries are reported to base their classification on topsoil properties alone. Our findings suggest that Bangladeshi farmers focus their interest on the topsoil/cultivated layer when asked to identify the soil on a plot. Our ethnographic enquiries into the soils of the Charan region produced no evidence that people identify and name soils as profiles Sillitoe 2000b – that is take several *vaj* layers in a series and give them an identifier/name (except for the *kasmati* category). The implication is not that they have no regard for the subsoil, neither do these comments imply that their understanding of soil is unsystematic. But one consequence is that local farmers and soil surveyors are basing their classifications on different pedological units, one focusing on the cultivated topsoil and the other on the soil profile including subsoil horizons. We should anticipate some incongruence, as reflected in the dissimilar maps of soil distribution.

GIS interpolation of local soil map: fact or fiction?

Furthermore the indigenous soil map created by GIS manipulation of the isolated plot soil data is an artefact unlikely faithfully to represent local knowledge.

Examination of the coincidence results suggests that the cartographic method is flawed. The reliability of an indigenous soil map produced through interpolation of sparse point data depends on three factors, the consistency of the local data collected, the density or coverage of those data, and the method used to interpolate those data to create a map.

The evidence suggests that farmers differ in their appraisal of soils, neighbours even classifying the soils of adjacent paddies differently (Figure 10.2). This patchiness reflects the contextual and relative character of local soil classification (Sikana 1993). Farmers are most familiar with the soils they encounter on their paddies, and they may vary in their experiences, as discussed earlier, such that one farmer's *atel* may not be the same as another farmer's *atel*. The methodology disconnects the ethnographic information from its owners and their social and cultural context, running foul of fundamental anthropological principles, treating local soil knowledge as immutable data that we can manipulate mathematically (Kloppenburg 1991). The GIS interpolation task was overly exacting, in attempting to delineate local soil units from plot data that lacked consistency. This disproves the proposition in Payton *et al.*, 2003), based on a preliminary analysis of the Bangladesh data, that 'the low earthen walls (*ails*) around paddies that are shown in the cadastral maps in Bangladesh obviate many of the soil unit boundary issues that arise from an open landscape' and that we might compile an indigenous knowledge soil map 'as an additive mosaic of plot-wise soil units'. Furthermore we used a low density of known data points to interpolate the map (15 per cent of plots having a local classification). The variable distribution of these data suggests that while increasing the sampling would affect the interpolated distribution it would neither change the chaotic pattern nor the mismatch with the surveyed soil map.

The interpolation procedure is dubious too. Soil surveyors are familiar with the problems of (i) interpolating between point data, usually obtained from augering or soil pits during soil survey; (ii) extrapolating this information to larger areas, and related to these; (iii) locating the boundaries between different soil units (Burrough *et al.*, 1997). Expert knowledge of soil-forming processes and field experience enables them to produce maps from low densities of survey points. It is more difficult to use indigenous soil data in this way, obtained from individuals with intimate knowledge of their plots but not necessarily beyond them. The soil surveyor may use expert scientific knowledge of soils to interpolate from a small number of sample points, using a universal classification system not directly formulated for the study site. We cannot interpolate highly mutable local knowledge of soils in this way, with its classification system highly specific to location, as the proximity analysis here attempts. Proximity analysis operates on a grid, starting with squares of known attribute – here, the qualitative, categorical attribute 'indigenous soil name'. It uses Euclidean distance to allocate identities (soil type) to blank squares based on closest proximity to squares of known attribute (Figure 10.4) (ESRI 1999).

The assumption that farmers' spatial knowledge accords to a Euclidean

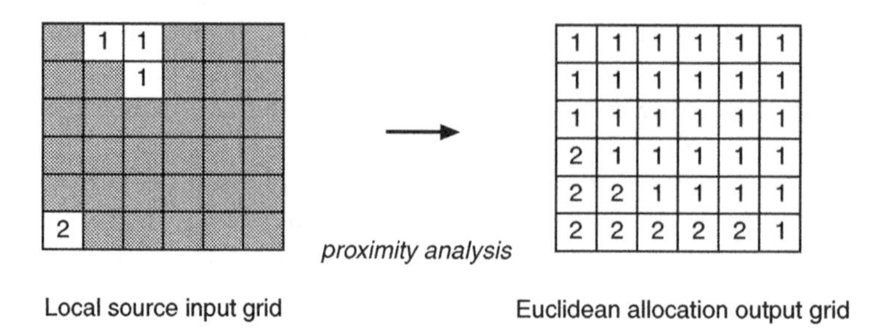

Figure 10.4 Proximity analysis.

metric is questionable. A review by Montello (1992) indicates that spatial cognition of the environment does not accord to a unitary spatial format, but is 'distorted, asymmetric, discontinuous and imperfectly co-ordinated'. Furthermore, the interpolation, although based on a soil survey comparison, used a dumb mathematical algorithm, with none of a soil scientists' (or farmers') expert knowledge of how depositional and landscape factors influence the distribution of soils, nor any sophisticated geo-statistical procedures such as kriging (Burrough 1997). It is possible to incorporate other farmer knowledge, such as slope, into GIS (Gonzales 1995) and make the algorithm smarter, processing soil information using these attributes. It is clear from their comments on the distribution of soils that Charan farmers have an awareness of soil forming processes, of the sort that informs many scientific soil classification schemes structured around the genesis of soils (Sillitoe 2000b). They are keenly aware of variations (*parthakkaya*) in soil across any area of land and that the soil environment is dynamic and subject to constant change, commenting for example on how clayey topsoil is deposited over sand below during floods. They not only observe these dynamic processes but also try to manipulate them, influencing topography and soil distribution to affect the productivity of their paddies.

The Charan region has had a complex sedimentation history that people can relate to the soils they see. Some 200 to 300 years ago there were only rivers with no people living in the area – according to some it was the Brahmaputra River. Oral history says that there were two ports at nearby Kalihati and Moricha (on the edge of the Modhipur Tract) with boats plying the water in between. At this time *char* land (river bars) *balu* 'sands' were deposited here. They say of *balu* that it is *mati janma* (literally soil birth), that is the first soil here, brought by the river long ago. Then other layers (*vaj*) have been deposited by floodwaters. The river gradually changed its course (people are well aware that rivers change their course over time, that the landscape is very dynamic). At first the land remained near to the flooding river and continued to receive heavy *ret* (silt) deposition. As the river moved further away, so the *ret* deposition declined, and the movement of clay increased, the *kumaira mati* clay gradually washed into the lower area from the higher surrounding land (the fine particles carried by the flood waters, leaving the coarser *balu* sand particles behind). When the river changed course, people started to settle the region and use the emergent land – the name Charan means grazing land, indicating the early use of the area for animals (together with fishing), by Hindus, the Muslims arriving later. The ***kumaira mati*** clay deposited on the surface changed into *atel* clayey topsoil with cultivation and the addition of organic matter.

These observations suggest that there are good grounds for correlating the soils identified in the pedological survey with those named by local people. But the distribution of soils in the Charan region is highly complex (Figure 10.1) and difficult to predict. In other areas, such as a site on the Ganges floodplain (McGlynn and Payton 1998b), the soils show a catenary relationship around a shallow *beel* basin. Farmers here discussed soils in relation to landscape position, distance above and away from the *beel* waterbody (Ghosh 2002), such that interpolation from local soil knowledge may be more reliable.

There is another reason why the two soil maps may not correspond closely. The survey and mapping procedure used by the soil scientists followed the standard double crisp approach to produce a cloropleth map. In this procedure, surveyors allocate soils to 'non-overlapping hierarchical soil classes linked to homogenous areas of geographical space' (Burrough *et al.* 1997). The crispness derives from: (i) delineating soil units with crisp boundaries, and (ii) assuming that the soil series so delineated are 70–80 per cent pure, that is one can allocate any given soil with a high degree of certainty to a single class in the classification hierarchy. As pointed out, soils occur in continua, grading from one to another through geographical space.[9] Double crispness is a pragmatic approximation used by soil surveyors. We are comparing the local soil map to a scientific one that may itself be a flawed representation of reality.

To map or not to map?

While farmers' understanding of soils suggests that there ought to be some correlation between indigenous and scientific soil classes, the interpolation method used to create the indigenous soil map fails to capitalise on it, and comparison with a double crisp soil survey map may be inappropriate. The attempt to use local knowledge to compile a map comparable to a soil survey may not have proceeded as anticipated but it teaches us some useful lessons and suggests future research avenues. It does not discount the use of GIS as an integration domain for scientific and indigenous soil knowledge, as a method for analysis of such spatial soil information. Two avenues suggest themselves; a high cost and low cost approach.

The expensive approach is untested and high risk, and fails to meet the original objective of a cheap methodology for collecting spatial information on soil resources, especially over large areas, relevant to the demands of sustainable development. It involves combining (i) a better understanding of how farmers classify their soils, and the criteria they use, employing in-depth ethnographic methods such as the observation and sorting methods described by Sillitoe (1996) and Furbee (1989) respectively, (ii) some form of survey carried out by farmers with indigenous knowledge researchers, and (iii) use of multivariate geo-statistical and fuzzy logic procedures (Bradley 1983; Burrough *et al.* 1997). The experiential dimension of local soil knowledge and classification results in farmers having variable knowledge based on differing areas. We need research into a methodology for investigating indigenous soil knowledge that can represent the complexity of mutable classification systems and variable subjective views on soil distribution.

The cheaper option is business as usual: the production of maps through facilitated dialogue between soil scientists and farmers, in a joint enterprise featuring current participatory methods. The dialogue will focus on evolving sketch maps. Numerous reports on participatory rural appraisal demonstrate that rural people can draw maps of the resources important to them, including soils (Furbee 1989; Birmingham 1998). This is relatively straightforward where the resources are discrete entities with mappable boundaries, such as forests, ponds and arable land. Soil maps are more problematic approximations given continuity in space. The experiential knowledge problems remain. These may be either mosaic maps of farmers' sketches of soil distribution on their land, or communally drawn aggregate maps (Payton *et al.*, 2003). A communally drawn map that is the product of discussion and probing, that carefully avoids a vocal few dominating the process, facilitated by either soil scientists aware of the social science milieu or anthropologists with an awareness of soil science, is likely to produce the best map.

Whichever avenue is followed, we commend the method employed here with those skilled in indigenous knowledge enquiries interacting with farmers initially to learn about soils and their distribution, and as necessary acting as facilitators in interactions with soil scientists. The soil surveyors should work independently to make their own reconnaissance investigation of the soils (Payton *et al.*, 2003). The two parties might then come together to negotiate a joint map, possibly using participatory GIS technology.

A number of caveats will apply to any such maps, and particularly the use of GIS, both their cartographic and cultural faithfulness. Studies of the spatial dimensions of knowledge indicate that it is not value neutral, as understanding of space is socially constructed. Those whose knowledge systems we characterize as local, experiential and subjective configure space through their own unique social and ecological history (O'Hanlon and Hirsch 1995; Raedeke and Rikoon, 1997). Consequently, as Couclelis (1992) cautions 'scientific representations of geographic space, and GIS in particular, cannot and should not try to mimic spatial cognition'. We must deploy the powerful analytical capabilities of GIS with care. Other critics make a similar point when they argue that GIS represent natural resources in a Western individualistic way, focusing on private property occupying discrete, non-overlapping space (Campbell

2002). GIS may misrepresent traditional land tenure, usufruct rights, and pastoral grazing institutions that are complex after their own social and cultural manner. The representation of indigenous soil knowledge using GIS, and maps generally, is equally problematic where classifications and understandings are likewise culturally, experientially and historically relative and contextual (Sikana 1993; Sillitoe 1996; Talawar and Rhoades 1998; Winkler Prins 1999).

Regardless of these cross-cultural conundrums, we have to recognize that maps often feature in natural resources development contexts. Development Interventions are by definition invasions of one socio-political order by another, in the belief that the scientific knowledge of one might help the poor. In these contexts we should strive to make the maps more relevant to local people caught up in interventions, to facilitate their participation. While the research exercise reported here became distant from the field, it had a practical aim – to produce cheap and relevant soils information to support sustainable development. In respect of evolving appropriate methodologies, we recollect Robert Chambers' concepts of optimal ignorance and appropriate imprecision (Chambers 1985).

Acknowledgements

This research is an output from research projects R6744 and R6756 funded by the UK Department for International Development in Bangladesh for the benefit of developing countries. The views expressed are those of the authors and not necessarily DFID. We wish to thank all research partners and local farmers for their invaluable contributions. Particular thanks are extended to Nitai and Dulu of Charan, Alice McGlynn and Robert Payton of Newcastle University's Centre for Land Use and Water Resources Research for work on the soil survey, P.-J. Dixon of Durham University's Anthropology Department for anthropological assistance, and Karl Pedersen of Durham University's Information Technology Service for help with GIS input.

Notes

1 It is at small scale that indigenous knowledge comes into its own.
2 Cadastral maps of the subcontinent date back to the commencement of the Survey of India's Revenue Survey in 1897. By the start of the twentieth century, much of India's cultivated land had cadastral maps at 1:3,960 scale, a mammoth achievement contained in 20,000 volumes (Kain and Baigent 1992). These maps have been periodically updated as land division continues. The cadastre contains an inventory of land parcels according to size, shape and location (Dale 1976). The function of cadastral survey is to establish the boundaries of land parcels and ownership for revenue collection through land tax.
3 The person directing the daily interviews had no prior experience of natural resources research nor soil science, and did not come from a farming family; he was essentially naïve. This overcomes Niemeijer's (1995) criticism that all interview-based approaches to elicit indigenous classifications (e.g. of soil) result in an etic taxonomy because the scientific paradigm influences interviewing consciously or unconsciously.

4 This agrees with the finding of Sinclair *et al.* (2000) that natural phenomena, such as the effects of certain types of trees and shrubs on soil conditions, are widely observed by, and known to farmers.

5 Van der Ploeg (1989), working with Andean farmers cultivating potato, concludes that 'personal knowledge of the field' is a significant component of local knowledge.

6 This very targeted indigenous knowledge survey of soils knowledge supplemented a much longer term ethnographic study of natural resource use at Charan *beel* (Barr 1998; Alam 2001; Payton *et al.*, 2003). Much of the information collected in the ethnographic survey was transcribed on to a word processor and collated in a computer-assisted qualitative data analysis software (CAQDAS) package. These interview transcripts were then sorted and coded in the package, enabling the researchers to bring together all the parts of different interviews where farmers discussed a particular soil type (Barr and Sillitoe 2000; Payton *et al.*, 2003). Interrogation of this coded interview database provided all the researchers with an understanding of the soils mentioned by farmers in the indigenous knowledge mapping exercise.

7 It is possible that variation in knowledge about land and soil resources could evidence some socio-cultural structuring. For example a sharecropper's knowledge might be different to a landowner's, a new landowner's knowledge [e.g. someone who has recently purchased land] might be different to a long established landowner [e.g. someone who has inherited land], the owner of a deep tube well might have a different understanding to other farmers, and so on.

8 GIS use two approaches to representing the world. In a vector model, they divide space into polygons of measured size and known location to which various properties, such as owner, soil type, elevation can be attributed. In a raster, or tessellation, model, they divide space into a regular grid of cells, each of which is characterized by the area it covers and other non-spatial properties of the cell (Lawas and Luning 1996).

9 Soil science puts considerable effort into developing more realistic representations of soil distribution in space and classification through geo-statistics that deal with the continuous nature of various soil properties in space, and fuzzy logic that deals with continuous classification (Burrough *et al.*).

References

Alam, M.A. 2001. Slaves of water: indigenous knowledge of fisheries on the floodplain of Bangladesh. PhD thesis, University of Durham.

Aldenderfer, M. and H.D.G. Maschner (eds). 1996. *Anthropology, Space and Geographic Information Systems.* Oxford: Oxford University Press.

Barr, J.J.F. 1998. Systems investigation of livelihood strategies and resource use patterns on Bangladesh floodplains. 994–1003. In *Proceedings of 'Rural Livelihoods, Empowerment and the Environment: Going beyond the farm boundary'*. 15th International Symposium of the Association for Farming Systems Research and Extension, 29th Nov–4th Dec 1998, Pretoria.

Barr, J.J.F. and J.G. Gowing. 1998. Rice production in floodplains: Issues for water management in Bangladesh. In *Water and the Environment. Innovative issues in irrigation and drainage.* (eds) L.S. Pereira and J.G. Gowing. London: E. and F.N. Spon 308–317.

Barr, J.J.F. and P. Sillitoe. 2000. Databases, indigenous knowledge and interdisciplinary research. In *Indigenous Knowledge Development in Bangladesh.* (ed.) P. Sillitoe. London: Intermediate Technology Publications and Dhaka: University Press Ltd. 179–195.

Berlin, B. 1992. *Ethnobiological Classification: Principles of categorization of plants and animals in traditional societies.* Princeton: Princeton University Press.

Berlin, B. and P. Kay. 1969. *Basic Color Terms.* Berkeley: University of California Press.

Birmingham, D.M. 1998. Learning local knowledge of soils: a focus on methodology. *Indigenous Knowledge and Development Monitor* 6 (2): 7–10.

Bradley, P.N. 1983. *Peasants, Soils and Classification. An investigation into a vernacular soil typology from the Guidimala of Mauritania.* Research Series No. 14. Department of Geography, University of Newcastle upon Tyne.

Brinkman, R. 1967. *Reconnaissance Soil Survey of Tangail District.* Directorate of Soil Survey of East Pakistan, Dhaka.

Burrough, P.A., P.F.M. van Gaans and R. Hootsmans. 1997. Continuous classification in soil survey: Spatial correlation, confusion and boundaries. *Geoderma* 77: 115–135.

Campbell, J. 2002. Interdisciplinary research and GIS: Why local and indigenous knowledge is discounted. In *Participating in Development: Approaches to indigenous knowledge.* (eds) P. Sillitoe, A. Bicker and J. Pottier. London: Routledge. Association of Social Anthropologists Volume.

Chambers, R. 1985. Shortcut methods of gathering social information for rural development projects. In *Putting People First.* (ed.) M. Cernea, Oxford: Oxford University Press.

Couclelis, H. 1992. People manipulate objects (but cultivate fields): Beyond the raster-vector debate in GIS. In *Theories and Methods of Spatio-Temporal Reasoning in Geographic Space.* (eds) A.U. Frank, I. Campari and U. Formentini. Lecture Notes in Computer Science 639. Berlin: Springer-Verlag. 65–77.

Dale, P.F. 1976. *Cadastral Surveys Within the Commonwealth.* London: HMSO.

Dent, D. and A. Young. 1981. *Soil Survey and Land Evaluation.* London: Chapman and Hall.

Dunn, C., P. Atkins and J. Townsend. 1997. GIS for Development: A contradiction in terms? *Area* 29 (2): 151–159.

EGIS and Delft Hydraulics. 1997. *Floodplain Fish Habitat Study.* Dhaka: Ministry of Water Resources, Water Resources Planning Organisation.

ESRI. 1999. *Arcview GIS 3.2 Helpfile.* Redlands, CA: Environmental Systems Research Institute.

Everitt, B.S. 1992. *The Analysis of Contingency Tables.* 2nd Edition. London: Chapman and Hall.

FAO-ISRIC-ISSS. 1998. *World Reference Base for Soil Resources.* World Soil Resources Report 84. Rome: FAO.

FAO and UNDP. 1988. Land resources appraisal of Bangladesh for agricultural development. Report No. 2 Agro-ecological zones of Bangladesh. Technical Report No. 5. Rome: FAO.

Furbee, L. 1989. A folk expert system: Soil classification in the Colca Valley, Peru. *Anthropological Quarterly* 62 (2): 82–102.

Gobin, A., P. Campling, J. Deckers and J. Feyen. 2000. Integrated toposequence analysis to combine local and scientific knowledge systems. *Geoderma* 97: 103–123.

Gonzales, R.M. 1995. KBS, GIS and documenting indigenous knowledge. *Indigenous Knowledge and Development Monitor* 3 (1).

Ghosh, G.P. 2002. Indigenous knowledge, livelihood and decision-making strategies of floodplain farmers in Bangladesh. PhD thesis, University of Durham.

Habarurema, E. and K.G. Steiner. 1997. Soil suitability classification by farmers in southern Rwanda. *Geoderma* 75. 75–87.

Harmsworth, G. 1999. Indigenous values and GIS: A method and a framework. *Indigenous Knowledge and Development Monitor* 6 (3): 3–7.

Kain, R.J.P. and E. Baigent. 1992. *The Cadastral Map in the Service of the State.* Chicago: University of Chicago Press.

Kloppenburg, J. Jr. 1991. Social theory and the de/reconstruction of agricultural science: Local knowledge for an alternative agriculture. *Rural Sociology* 56 (4): 519–548.

Kundiri, A.M., M.G. Jarvis and P. Bullock. 1997. Traditional soils and land appraisal on Fadama lands on northeast Nigeria. *Soil Use and Management* 13: 205–208.

Lakoff, G. 1987. *Women, Fire, and Dangerous Things: What categories reveal about the mind.* Chicago: Chicago University Press.

Lawas, C.M. and H.A. Luning. 1996. Farmers' knowledge and GIS. *Indigenous Knowledge and Development Monitor* 3.

McGlynn, A.A. and R.W. Payton 1998a. *Soil Survey Report, Charan Beel, Tangail District, Dhaka, Bangladesh.* Newcastle: Department of Agricultural and Environmental Science, University of Newcastle upon Tyne.

—— 1998b. *Soil Survey Report, Ujankhlasi, Rajshahi District, Bangladesh.* Department of Agricultural and Environmental Science, University of Newcastle upon Tyne.

Mohamed. M.A. and S.J. Ventura. 2000. Use of geomatics for mapping and documenting indigenous tenure systems. *Society and Natural Resources* 13: 223–236.

Montello, D.R. 1992. The geometry of environmental knowledge. In *Theories and Methods of Spatio-Temporal Reasoning in Geographic Space.* (eds) A.U. Frank, I. Campari and U. Formentini. Lecture Notes in Computer Science No. 639. Springer-Verlag. 136–152.

Neimeijer, D. 1995. Indigenous soil classifications: Complications and considerations. *Indigenous Knowledge and Development Monitor* 3 (1): 20–21.

NRI. 1999. *Report on a Workshop on Geographical Information Systems and Participatory Methods.* Held at Commissioners House, Historic Dockyard, Chatham; 7 July 1999. Natural Resources Institute, University of Greenwich, Kent.

Office of Field Services. 1993. *Agroecological Zones at Thana Level.* Office of Field Services, Department of Agricultural Extension, Dhaka.

O'Hanlon, M. and E. Hirsch (eds). 1995. *The Anthropology of Landscape: Perspectives on place and space.* Oxford: Oxford University Press.

Payton, R.W., J.J.F. Barr, A. Martin, P. Sillitoe, J.F. Deckers, J.W. Gowing, N. Hatibu, S.B. Naseem, M. Tenywa and M.I. Zuberi. (2003). Methodological lessons from contrasting approaches to integrating indigenous knowledge and scientific soil and land resources survey. *Geoderma* 111 (2): 355–386

Pickles, J. (ed.). 1995. *Ground Truth: The social and economic implications of geographical information systems.* New York: Guildford Press.

Ploeg, J.D. van der. 1989. Knowledge systems, metaphor and interface: The case of potatoes in the Peruvian Highlands. In *Encounters at the Interface: A perspective on social discontinuities in rural development.* (ed.) N. Long. Wageningen Studies in Sociology, 27. Wageningen: University of Wageningen.

Raedeke, A.H. and J.S. Rikoon. 1997. Temporal and spatial dimensions of knowledge: Implications for sustainable agriculture. *Agriculture and Human Values* 14: 145–158.

Rosch, E. 1975. Universals and cultural specifics in human categorization. In *Cross-Cultural Perspectives on Learning.* (eds) R.W. Brislin, R.W. Boehner and W.J. Lonner. New York: Wiley. 177–206.

Sillitoe, P. 1996. *A Place against Time: Land and environment in the Papua New Guinea Highlands*. Amsterdam: Harwood Academic.
—— 1998. Knowing the land: Soil and land resource evaluation and indigenous knowledge. *Soil Use and Management* 14: 188–193.
—— (ed.) 2000a. *Indigenous Knowledge Development in Bangladesh*. London: Intermediate Technology Publications; Dhaka: University Press Ltd.
—— 2000b. Cultivating indigenous knowledge on Bangladeshi soil: An essay in definition. In *Indigenous Knowledge Development in Bangladesh: Present and future*. (ed.) P. Sillitoe. London: Intermediate Technology Publications; Dhaka: University Press Ltd. 145–160.
—— 2002. Contested knowledge, contingent classification: Animals in the highlands of Papua New Guinea. *American Anthropologist* 104 (4): 1162–1171.
Sikana, P. 1993. Mismatched models: How farmers and scientists see soils. *ILEA Newsletter* 9 (1): 15–16.
Sinclair, F.L., D.H. Walker, B. Thapa, L. Joshi, P. Preechapanya, and A.J. Southern. 2000. *General Patterns in Indigenous Ecological Knowledge*. Paper presented at Association of Social Anthropologists' Conference 2000.
Tabor, J.A. 1992. Ethnopedological surveys – Soil surveys that incorporate local systems of land classification. *Soil Survey Horizons*. Spring, 1–5.
Tabor, J.A. and C. Hutchinson. 1994. Using indigenous knowledge, remote sensing and GIS for sustainable development. *Indigenous Knowledge and Development Monitor* 2 (1): 2–6.
Talawar, S. and R.E. Rhoades. 1998. Scientific and local classification of soils. *Agriculture and Human Values* 15: 3–14.
Weiner, D., Warner, T., Harris, T. and R. Levin. 1995. Apartheid representations in a digital landscape: GIS, remote sensing and local knowledge in Kiepersol, South Africa. *Cartography and G.I.S.* 22 (1): 30–44.
Winkler Prins, A.M.G.A. 1999. Local Soil Knowledge: A tool for sustainable land management. *Society and Natural Resources* 12: 151–161.

Chapter 11

Keeping tradition in good repair

The evolution of indigenous knowledge and the dilemma of development among pastoralists

Paul Spencer

This chapter is concerned with the arid region associated with nomadic non-Islamic pastoralists in East Africa and refers to the period before the penetration of the cash economy and the process of globalisation. The exclusion of Islamic pastoralists living beyond the northern perimeter of this region is significant. Islam spread into Africa along trade routes, and these skirted the region rather than passing through it, because of its rough and arid terrain. It has been suggested that it was the effectiveness of warrior age organization among these pastoralists that checked the spread of Islam. However, a more likely explanation is the sheer absence of long distant trades routes through the region.[1] To this extent, indigenous knowledge was less likely to be infiltrated by ideas that stemmed from expanding civilizations in earlier times.

Among these pastoralists, knowledge of their herds was nurtured within each corporate family. The family was the unit of production, and was normally under the authority of the most senior male. It was through families that wealth accumulated and passed down the generations; and I have argued elsewhere that East African age systems have to be viewed with this in mind. It is no distortion to regard pastoralism in this region as a family enterprise to which all members were committed, or they faced being squeezed out of the pastoral niche.[2]

At a more inclusive level than the family, the term 'tribe' was particularly apt when applied to pastoralists, for this conjures up the image of a bounded social entity. Nomadism tended to create cultural uniformity over a wider area as families migrated with their stock independently of one another. From the stock-owner's point of view, his community of reference extended to wherever he happened to be, even if his neighbours changed with every nomadic movement. It follows that it was this transient community who represented the 'tribe' as repositors of tribal custom at any local meetings, sharing a much wider experience. This uniformity within each tribe corresponded to sharp intertribal boundaries that separated neighbouring ethnic groups culturally and linguistically. To the extent that intermigration and intermarriage did not occur on any significant scale across these boundaries, indigenous knowledge among the nomads was not shared with these neighbours. In this way, the pastoralists contrasted with settled agricultural groups, where there was often local variation and even a blurring of identities and dialects along the boundaries.

A question of resilience

A popular view of nomadic pastoralism argues that they struggle at the rough end of a balance with nature, where they are closer to the environmental forces that shape their decisions than Western advisers with only a partial knowledge. As one ecologist expressed it: '. . . traditional nomadic society was approximately in equilibrium with natural resources on which they entirely depended; any mistake in land use was penalised by reduction of the land carrying capacity for human and animal populations. Hence, they had to learn sound land-use practices in order to survive'.[3] This assumes that the development of indigenous knowledge by trial and error was on a par with the adaptation of savannah ecosystems over millions of years prior to human habitation. Other writers have emphasized the virtue of nomadic adaptability as a way of life that has surmounted periods of regional instability and fluctuations of climate or tsetse infestation, well above the threshold of environmental degradation.[4] This assumes some protective hidden hand that has an affinity with the functionalist approach that is well represented in pastoralist studies. These too emphasize the robustness of tradition, irrespective of the inevitability of change. Pastoralist societies are, after all, remarkably resilient, and this resilience deserves explanation.

The harsher side of this approach notes the Malthusian downside of the proverbial wealth of pastoral peoples and the extent to which there has always been a 'sloughing off of poor herders from Africa's pastoral sector'.[5] Natural selection took its toll and the survivors were those who were best adapted to their sparse environment. A useful example of this process and of the importance of bonds extending beyond the family is provided by the Turkana, who inhabit a particularly bleak area in northern Kenya and rely heavily on mutual help. This is underpinned by the regular exchange of stock within a network of stock associates, who are frequently also affines and best friends. Over the years, these exchanges build up trust between stock owners. But mutual help alone cannot cope with the increase in population, and this has prompted aid agencies to set up refugee camps, attracting Turkana families who have lost their stock. To reduce dependency among refugees still committed to pastoralism, these agencies have offered selected families the nucleus of a new herd to return to the nomadic economy. A study of these by Vigdis Broch-Due has revealed that in losing their stock and looking elsewhere for support, these refugees had broken their network of nomadic associates and lost their trust. They needed to rebuild this trust to re-establish themselves within the exchange economy. But they could not achieve this quickly enough to survive the hazards of their environment unaided, and this precipitated their return to the refugee camp as paupers once more.[6] Again, one may note the aptness of the model of a family enterprise that can only recover from bankruptcy by surmounting a wider crisis of confidence.

An even bleaker view is neo-Malthusian, emphasizing the irreversible damage that pastoralists inflict on the land through overgrazing. This does not question the resilience of some of these societies but draws attention to the limited resilience of their life-support system. Approaches to the development of pastoralism have assumed that either they should be encouraged to settle as agro-pastoralists or at least be confined to defined areas within the limits of a sustainable eco-system. I will return to these diverging models after elaborating on the pattern of adaptation and the role of indigenous knowledge within this pattern.

The dynamics of diversification

Clearly, the harsh conditions seriously curtailed survival rates in earlier times. However, there were alternative niches to which refugees could turn, and inter-tribal migration is a frequent theme in oral histories, sometimes along established paths and revealing a form of adaptation within the region at large. This was a two-way process. At times, the pastoral niche absorbed a trickle of hunter-gatherers, agriculturalists, and displaced pastoralist refugees from elsewhere, and the flow would be in the opposite direction at other times. But the route into pastoralism tended to be as hired herders who only managed to build up their own herds against the odds.[7] Intermigration clearly sharpened the general awareness of other peoples, and migrants would keep certain ritual practices associated with their family or clan; but in discarding old niches and adapting to new, these external movements only appear to have led to a wider sharing of indigenous knowledge in exceptional cases.

One of these exceptions were the Chamus, and though atypical, they provide a striking illustration of the process of adaptation before, during and since the colonial era. The Chamus were established close to Lake Baringo in the Rift Valley of Kenya, where they had their own age organization. They were surrounded by pastoralist peoples, including the Maasai and Samburu to whom they were distantly related as fellow Maa-speakers. Chamus oral traditions suggest that they underwent a series of economic transformations. Taking these as an authentic reflection of history, whatever the distortions in matters of detail, each innovation is presented as an opportunity that spread as a new and increasingly dominant idea, displacing earlier traditions. This series of transformations may be examined in Darwinian terms of adaptation through a process of selection, rather as Warwick Bray has outlined with reference to the archaeological evidence of social change in South America.[8] In other words, Darwin's biological model provides a metaphor, referring to shifts in Chamus culture – in knowledge that informed their way of life – rather than to the evolution of their human population as a species.

The Chamus claim that they had originally been hunter–gatherers; and their transition to agriculture is portrayed as a chance event by serendipity (cf. 'mutation'): an elder picked up a sprig of finger millet that had been dropped by a migrating bird, decided to plant it and then gathered the first crop. He gave seed to other Chamus and this new source of food spread among them ('adaptive radiation'). Over a prolonged period, the Chamus then developed an intricate system of family-based irrigation, supervised by a council of elders ('stabilization').

By the mid-nineteenth century, this provided a food surplus, enabling them to accommodate refugees from surrounding peoples. Foremost among these were some pastoralists who had become detached from the main body of Samburu and had developed their own practices including fishing ('genetic drift'), and then they lost all their stock. However, this was not the 'extinction' of a way of life, for they still had their pastoral and fishing skills; and when they migrated to join the Chamus, they introduced these skills and rebuilt their herds in return for their adoption into the irrigation system. They were the principal agents in the transition of the Chamus to agro-pastoralism ('hybridization'). Meanwhile, the prior existence of the council of elders was fortuitous in coordinating this transformation of Chamus society ('preadaptation'). At about this time, coastal caravans were beginning to penetrate the area; and the Chamus increased their irrigation production to meet the growing opportunity to exchange food for goods ('specialization'). The system became overworked, flagged, and then the major part was destroyed by a flash flood in 1917. This was due to over-exploitation according to one authority or overgrazing according to another ('overspecialization' threatening 'extinction'). Either way, pastoralism had become the most successful component of their mixed economy, and this was consolidated by adopting the Maasai system of warrior villages to guard their cattle ('selective adaptation').[9]

Most recently, the community basis of Chamus agro-pastoralism has been undermined by the transition to the fringes of the capitalist economy in post-colonial Kenya ('anagenesis'). New ideas and strategies for accumulating new forms of wealth have been divisive, creating an unbridgeable gap between rich and poor, and a rift between older traditionalists and the younger generation of opportunists ('cladogenesis'). Unlike the Turkana instance, it was those who tried to persist in the traditional system who were least successful, and their way of life seems destined for extinction.[10]

These adaptive features are clearly not unique to Chamus experience. They have incidental parallels with oral traditions elsewhere in the region, reflecting a certain historical flux between pastoralism, agriculture, foraging, and mixtures of these in the process of local adaptation. Most recently, the infiltration of the cash economy has affected them all.

Adaptability and discourse

This extended metaphor notes the Darwinian parallels. But without identifying the mechanism, it provides little more than a set of descriptive labels. In this context, one should note that each stage in the development of the Chamus economy was implemented by a corresponding development of understanding.

indigenous knowledge, in other words, underwent a constant process of renewal in response to changing circumstances.

The role of the elders in this process was generally associated with the various age systems in the region. Younger adventurous men could aspire to wealth in the short term through raiding (and nowadays as entrepreneurs), but it needed the experience of older men to care for longer-term interests. In other words, these age systems were not without their contradictions, but in times of peace, elders were the ultimate repositors of experience; and in times of turmoil, it was this experience and their diplomatic credentials that were a link with the future. Elders were respected for their wisdom. And greater still was the combined knowledge of such men that surmounted the partiality of individuals. So great was the reputation of older men that their wisdom and knowledge was sometimes held to be next only to God's.[11]

A key activity in the continuous process of adaptation was the debating among elders at their formal meetings to resolve immediate problems, interpreting or reinterpreting the nuances and relevance of tradition as the situation seemed to demand. There was a tacit acceptance of creeping change, but above all a premise that the wisdom of elderhood lay in pooling their experiences and insights in order to arrive at well-considered courses of action. The wisdom of tradition in coping with the unexpected was seen to lie in this community of knowledge and discourse. With a relatively simple material culture imposed by their hazardous environment, oratory may be considered as a creative artform in its own right in this region and it certainly impressed various early observers.[12]

The scale of oratory varied considerably. The most imposing debating arena in the region was in the Ethiopia-Kenya border area, where Booran age-sets entered the *gada* grade for successive periods of eight years. During each *gada* period, an array of office holders were nominated to take responsibility for resolving all forms of conflict and dispute; and midway, a massive pan-Booran assembly was mounted to consider intractable problems. No aspect of customary law was immune from scrutiny on these occasions, and the debating was geared towards updating tradition in order to adapt to the realities of change. An altogether more raw and localized form of democracy has been noted by Neville Dyson-Hudson among the Karimojong of northern Uganda, where any forceful elder could try to impose his views in a debate. Having emerged as a local leader, he might order persistent objectors to leave the meeting, provoking a minority to vote with their feet and to form a rival faction with their own spokesman and debate. If this movement threatened to wrest the initiative, the local leader would be forced to climb down and reassess his views in order to remain within the mainstream of discussion. The thrust towards consensus overrode the clash of personalities.[13]

A particularly sensitive analysis of elders' debating has been provided by David Turton in relation to the Mursi of southern Ethiopia. The most influential Mursi elders were those with a flare for piecing together a forceful argument that assimilated different points into some imaginative synthesis, overriding parochial interests. Such men had to cultivate their reputation, or popular regard would shift to those who outshone them, especially ambitious younger men. Each debate had the potential to modify the contours of influence, based on attendance and performance. Each speaker needed to cultivate his audience. He should only interject at a point when they were ready to listen, and then hold their attention with a terse style and subtle allusions that made a significant contribution towards the discussion. He should then finish at a point of his own choosing, before he was hassled by a rising tide of interruptions. If he attempted to intervene prematurely, when attention was still focused on another speaker, or too often, or with too little to say, then he would lose face. No individual was indispensable, and there was no specific person to arbitrate between conflicting views. Rather there was an implicit process of peer review, with the most influential contributions emerging towards the end, bringing together the strands of argument and reducing the need for further discussion.[14]

The Maa-speaking peoples followed a similar pattern. Any elder could speak at their debates, but had to obey the rules of procedure, taking his turn, addressing the issue, and making a coherent contribution or others would shout at him to sit down. Those who rose to the occasion would command the space around them with the deft use of their sticks and their timing, when even their extended pauses and repetitions held the audience. The Chamus council of elders, like their Samburu and Maasai neighbours, recognized the authority of debate, pooling their views when faced with a problem, until some consensus was reached that was binding on everyone. The Samburu compared a debate to the acacia tree in whose shade the elders would assemble: they all would come with their own points of view (the branches), and the discussion lasted until they had resolved their differences to arrive at a binding compromise (the trunk). A Maasai metaphor made a similar point, referring to the spokesman for any age-set as their 'head', while influential men who represented shades of opinion in discussion were his surrounding 'feathers', as in a warrior's headdress. The spokesman's skill was to bring together these diverse views in the course of debate, without declaring his hand too soon. He was expected to listen and then to steer the debate towards a binding consensus.[15]

A point to emphasize is that in the more casual discussions and gossip of daily life, men of influence would be priming themselves with relevant information that underpinned their performance on the more formal occasion. Among the Maasai, a pressure group might appoint a particularly adept member to lead any discussion on their behalf. He could not refuse this, and it gave him a certain authority to persist in asserting their case in debate. Among his peers and less formally, he was in effect their head, arriving with them at a sturdy consensus view. On the formal occasion subsequently, he was a feather, asserting this point of view in contention with others. This steered community decision-making, gleaning relevant knowledge and perspectives from all quarters. Rather than some kind of manual of indigenous knowledge that the most influential elders held in their heads, this aspect reveals 'tradition' as a broad tenet, a framework, and 'knowledge' as a realm of possibilities and points of view that were cultivated in the process of community life.

The egalitarian thrust of pastoral societies in East Africa provided the opportunity to pool creative strands of understanding through democratic discussion in their debates. The oratory of these occasions stimulated a process of selection by popular acclaim and credibility in order to arrive at some working consensus. Through performance, the collective decision would be binding in the first instance, and memorable in the longer term, having established itself in the collective memory.

The notion of a collective wisdom that is cultivated and harvested in debate brings this argument back to the Darwinian parallels that were noted earlier. This has been elaborated by Karl Popper, whose theory of the growth of scientific knowledge extends to cosmologies and inventiveness in general, entailing modes of performance, personality clashes, and peer review, and no presumption of progress.[16] In Popper's model, human awareness and endeavour focus on the problems of existence as they occur (cf. Heidegger); and people, as individuals and as groups, learn through trial and error. In this process, selective pressures weed out ideas and experiments that do not stand the test of reality. They either succeed and become incorporated into normal practice in a process of adaptation, or they fail and will be discarded or ignored. Either way, the body of knowledge adapts to the reality. Routine patterns of response build up through the experience of failed attempts (cf. Pavlov); and where a pattern establishes itself as an underlying working premise then this provides a basic understanding – a strategy that suffices until it too is put to the test.

At the community level, Popper envisaged a world of potential knowledge that has an autonomy of its own, lying beyond the awareness of any single knower. Through discourse, there is a spontaneous process of revelation and criticism as ideas jostle for attention. The fate of a novel idea, like a mutation, depends on success or failure. If it succeeds, it spreads and becomes incorporated into the body of knowledge. From this angle, it may be a moot point whether the innovation is a personal discovery (e.g. the first Chamus to plant finger millet), or borrowed (e.g. the diffusion of Samburu pastoralism into the Chamus economy or the adoption of the Maasai warrior village system), or pure chance (e.g. inspired by the random behaviour of a diviner's oracle). The significant point, as E.B. Tylor once noted, is that the community are ready to adopt it.[17]

Pursuing the Darwinian analogy, there is a cultural pool of awareness – the accumulated experience, imagination, and partial knowledge of individuals – that corresponds to a gene pool in biology, where each organism contains only a partial combination of the available genetic information. The coming together of minds on a particular problem, as occurs in formal or less formal discussions among pastoralists, selects from the assembled body of knowledge. The pool of ideas provides a fertile breeding ground for new combinations; and out of this potential experience, new propositions are thrown up that undergo a critical process of selection until one is favoured and put to the test (cf. the Samburu acacia tree).

The model can be elaborated with regard to the underlying premises that structure the syntax of knowledge and provide strategies for interpretation and action. When there is a radical shift in circumstance, survival may hinge on the community's ability to engage in a self-critical dialogue that breaks through the shell of their basic doctrines, and various imaginative 'mutant' forms may enter the mainstream of awareness, rewrapping the old package to form a new one (cf. Kuhn's paradigm shifts). In Popper's scheme, it is not society as a biological entity or species that is threatened with extinction by changing surroundings, so much as rigidly held cultural premises. Cultural regeneration is achieved through a selective process of creative social rather than procreative sexual discourse: the stuff of history rather than of genetic evolution.

From this point of view, the successive transitions of Chamus society were bound up with a discourse that was hammered out in their council of elders and arose out of events and ideas that lay beyond the knowledge or experience of any one member. Transitions of this magnitude are less characteristic of nomadic pastoralist societies in their oral traditions, but the significance of debate in the selective process of adaptation is quite explicit.

The pastoral community and the tragedy of the commons

indigenous knowledge and tradition among pastoralist communities in this region, then, appear to have been kept in good repair as they were constantly tested by the turn of events. This seems to support the benign view that their adaptation to their environment has been shaped by generations of experience. However, there remains the more pessimistic neo-Malthusian view that pastoralists overgraze their pastures, undermining their life support system. When Garet Hardin (1965) coined the expression, 'the tragedy of the commons', he took free-ranging pastoralism as his prime example. Where pastures are shared, he argued, no owner has an incentive to restrict the size of his herd in order to conserve this common land. If one of them tries, he would have no guarantee that others would do the same, putting his moderate herd at risk as the pasture is degraded. Hence his best strategy is to maximize the size of his herd in order to increase the chances of survival. In a free-for-all, altruism is self-defeating, and short-term personal gain overrides the long-term public interest. In this article, Hardin was citing unrestricted overgrazing by pastoralists as an allegory of human excess in general, and ultimately the survival of civilization itself, for the ecological crisis is world-wide. In the final resort, we are all caught up in a poverty-trap.[18]

The strength of pastoralist communities, associated with their emphasis on consensus as a widespread ideal, questions the aptness of this image of self-seeking herders. Pastoralists shared common interests and knowledge concerning the welfare of their stock, and displayed a form of democracy that constrained individuals, as outlined above. To this extent, the tragedy of the commons is inappropriate.

However, the basic problem raised by Hardin remains. The robustness of pastoralist cultures that places adaptability on the shoulders of shared conscious human experience does not address the broader historical issue regarding the damage caused by the unintended consequences of action. It excludes issues that are not seen as relevant to the longer-term interests of society at large. The problem concerns the limits of collective understanding among nomads.

This is well illustrated in a comprehensive survey of indigenous knowledge in developing countries (Warren *et al.* 1995). Of the 47 studies included in this work, only two are directly relevant to pastoralism. One of these is an essay on ethno-veterinary medicine, and this is packed with a detailed breakdown of the topic from a wide range of sources.[19] This array of data reflects what anthropologists have often claimed: that pastoralists lavish attention and care on their stock, and know them intimately – both as species with different needs, and as individual animals with their own personalities and foibles. Matters concerning the care of stock or trends that have a relevance for the well-being of herds are public knowledge, and any issue in discussion that involves stock is inevitably sensitive.

The other essay concerns the traditional management of (semi-) arid land, and this raises the serious issue of the commons.[20] The treatment in this essay provides a sweeping outline of various sources, but in assuming that pastoralists seek to conserve their land, the author evades the critical issues of indigenous understanding and usage. The superficial argument concerning control over resources appears true up to a point, but the supporting literature does not suggest that pastoralists regard the conservation of their environment as a vital issue, and this leaves the more fundamental issue unresolved.

If one considers indigenous knowledge concerning cattle management and ethno-veterinary medicine, then this may be benign and sound, stemming from the accumulated feedback of direct experience. But wisdom derived from the care of livestock is altogether more immediate than a wider ecological understanding. Nomadic pastoralists survived through their mobility, and when the grazing was exhausted, they moved on, taking their herds with them. As a result, there was little feedback (if any) concerning the recovery of the land they left behind. Taken together, these two essays suggest that the fund of pastoralist knowledge has built up around the care of livestock rather than of the land itself. Popular concern focuses on the short-term resilience of the herds and grazing rather than the long-term resilience of the top-soil. The process of experiencing the link between nomadic herding practices and the care of common land is more extended, and the benign argument is harder to sustain at this level.

The allegation that pastoralists undermine their life-support system has led to counter-claims that they once had systems of grazing rotation to conserve their land. A number of writers have suggested that the Maasai and Samburu, for instance, previously protected their land through indigenous forms of grazing control.[21] However, the evidence cited in these articles focuses entirely on short-term aspects of land management, and this leaves open the question of whether these Maa-speakers were aware of more fundamental ecological issues. This is not to question whether various local agreements within Maasai communities, coupled with the flexibility of their herding arrangements, were effective in conserving grazing up to a point, but these were essentially seasonal accommodations. Throughout the wider region, the issue of local consent was clearly crucial. Those who shared resources in any locality could combine to restrict the free use of limited supplies by outsiders. This often concerned access to water points and their maintenance (Maasai, Samburu, Booran, Jie, Karimojong), or access to pasture in situations of scarcity (Samburu, Chamus, Booran, Turkana), and especially by distant neighbours (Maasai, Samburu, Karimojong).[22] But land as a long-term sustainable resource was not a consideration.

During my own periods of fieldwork, the Maasai were hostile to government systems of grazing control to conserve the land. It was their resentment against imposed schemes on their land that they would stress, rather than any counter-argument that they traditionally imposed their own self-regulation towards the same end; and in the areas outside these schemes, they still claimed the right to free access to all pasture and water within their tribal territories. The Samburu had a similar attitude towards imposed schemes, and because they also had interspersed clans that were autonomous throughout the area, it would have been even harder to enforce elaborate means of controlling grazing.[23] Such patterns of grazing as existed were matters of individual preference and expedience rather than of prescription. Indigenous systems of grazing control to conserve the land itself appear to have been generally absent among nomadic pastoralists and would probably have been unworkable.[24] To this extent, the unintended consequences of their practices in the longer-term could have been more serious than they realized.

Put simply, when pastures were exhausted during a dry season, leading to decisions to migrate, irreversible damage to the land may already have been inflicted, and the cumulative effect of this damage may not have been self-evident. A dramatic illustration of the tragedy of the commons occurred in 1917, when a flash flood destroyed the principal irrigation system of the Chamus. An analysis of this event by Robert Chambers (1973: 346) has suggested that it was a direct result of heavy overgrazing that followed unrestrained growth in Chamus herds, undermining the natural drainage that fed the system. This was clearly not deliberate, and again it raises the question of the limits of popular awareness of ecological issues.

The argument that pastoralists traditionally conserved their land, then, may be questioned on the grounds of its irrelevance to their nomadic pattern of existence. However, access to land has become a critical issue among pastoralists generally, especially following land registration in Kenya. These peoples no longer have unlimited tracts over which to graze, and any further deterioration as their land erodes under the hoofs of their herds now becomes more immediately relevant. This provides grounds for a tentative optimism in the continuous process of updating knowledge against the depressing background of the demise of much else. A change in perception of land as a resource in an unusually arid region is illustrated again by the Turkana.

On a scale of pastoralist organization, ranging from the elaborate *gada* assemblies of the Booran to ad hoc gatherings precipitated by issues of immediate concern, the Turkana would be situated near the bottom end of this range. In 1950, Philip Gulliver noted that the Turkana had recently begun to acquire camels from the Rendille and treated them as cattle, using pasturage that was common to all; the mixed herd could be moved anywhere at any time. Since then, the extent to which access to grazing has led to disputes, discussions, and compromises to resolve confrontation, is an indication that land use has become a communal issue. By 1980, Frode Storas noted that families still owned mixed herds, but they had begun to specialize in different types of stock, recognizing that they were better suited to different types of environment and management. Each stock-owner continued to maintain a close network of stock associates, but whereas the earlier study had focused on the importance of regular exchanges between them to maintain mutual confidence in the event of loss, the later study noted the vital role of this network in acquiring exclusive access to pastures. Associates, who specialized in a certain kind of stock and had access to suitable grazing, would be entrusted with herding each other's camels or cattle or small stock, displaying once again a symbiotic confidence in one another. Attempts to bypass these exchange arrangements in response to immediate needs would lead to confrontations that might be settled by discussion or force. Whereas the earlier study had drawn attention to stock-ownership as the critical issue of contention within the family, the later study drew attention to the extent to which access to grazing was also now a critical community issue.[25]

Conclusion: the tragedy of globalization and the future of discourse?

This chapter has concerned indigenous knowledge among East African pastoralists in their traditional pre-monetary setting. This has been linked to the processes that sustained community life, adapting as circumstances changed. A significant change during the twentieth century has been the confinement of nomadic pastoralists to defined areas – at first to their tribal reserves and more recently into smaller units – and the conservation of their land has become a major issue. Unlike family-based skills in herding and the care of their animals, concern over their land is precisely the sort of knowledge that is a matter for the public arena.

This concern is, of course, widespread, but the dilemma facing pastoralists is particularly acute. As they become enmeshed in the margins of the expanding world economy, they are at the tail-end rather than at the source of the problem, aptly described as the tragedy of the commons. But seen from another point of view, it is the tragedy of globalization. New forms of wealth have undermined the corporate basis of community life, which has been the foundation upon which indigenous knowledge is constantly regenerated. The new forms of knowledge have by their very nature served the individual rather than the community. They have provided immense opportunities for a privileged few who are not (or are no longer) intimately involved in local community life. The most wealthy are extending their control over resources, increasing the pressure elsewhere. As less wealthy cultivators migrate to colonize marginal areas and as refugee camps develop into settled communities around permanent water points, pastoralists are pushed further still into the arid wastes beyond. Or they are caught up in a downward spiral where the survival of their families has been achieved by foraging the margins of civilization. This follows the earlier pattern of sloughing off the least successful pastoralists, but at an unprecedented pace to a point where it is pastoralism in its traditional sense that is sloughed from the mainstream of development. There is a sense in which this increasing polarization of wealth and opportunity echoes Durkheim's concern with anomie, which describes the disorientation of individuals at both ends of the social spectrum: those with unlimited means who lose direction, and those deprived of sufficient means to achieve any meaningful goal. The loss of community with all its constraints is the loss of the middle ground.

The emergence of a new form of society based on inequality and patronage appears to be gathering pace in this region, encroaching on the traditional autonomy of indigenous peoples. And I would argue that this is just what occurred in earlier centuries among pastoralists further north, as they were caught up in the spread of Islam.[26] However, this creeping process of civilization has not resolved the ecological dilemma anywhere. It distances those who have ultimate control over resources from the problems of sustainability.

In the face of increasing alienation of land in this region, pastoralists are seeking guarantees for the security of their tenure.[27] This is couched in terms of a guaranteed basis for maintaining their livelihood. But to this, one may add that it is the best hope for preserving community life and thereby cultivating a public discourse – an indigenous knowledge – concerning the future of their environment.

Notes

1 Trimingham 1968: 37–39, 128; cf. Johnston 1886: 408.
2 Spencer 1998: 26–34.
3 Houerou 1986: 140; cf. Spooner 1971; Baxter 1993: 157.
4 Baier 1976: 5; Kjekshus 1977: 67–68; Swift 1977b: 457; Homewood and Rodgers 1987: 122–124; Gowlett 1988: 44.
5 Little 1992: 115.
6 Gulliver 1955: 196–203; Broch-Due 1990: 147–154.
7 Spencer 1973: 199–209; 1998: 267–269; Berntsen 1979: 110–114; Schlee 1989: 145–236.

8 Bray 1973.
9 Chambers 1973: 346; Sobania 1980: 62, 77; Anderson 1989: 88, 93–95; Spencer 1998: 139–145, 151–157.
10 Little 1985a; 1992: 91–104.
11 Dyson-Hudson 1966: 181, 212; Lamphear 1976: 153–155, 246; Spencer 1988: 49.
12 E.g. Among the Maasai, Thomson 1885: 162; Hinde 1901: 33; Merker 1910: 86; Jackson 1930: 294.
13 Dyson-Hudson 1966: 223–224; Legesse 1973: 81–98.
14 Turton 1975: 171–178.
15 Gulliver 1963: 61–64, 224–231; Spencer 1965: 176–177, 180–184; 1988: 105, 215.
16 Popper 1963: 216–222; 1973; 1977.
17 Tylor 1865: 378–379.
18 Hardin 1968: 1244; cf. Royal Commission on East Africa 1955: 294; Livingstone 1977: 210–220; Dyson-Hudson, N. and P. 1982: 234–5.
19 Mathias-Mundy and McCorkle 1995: 488–498.
20 Niamir 1995: 245–257.
21 Jacobs 1975: 417; 1980: 287; Galaty 1980: 164; Ndagala 1990: 177–178; Sperling and Galaty 1990: 80; cf. Niamir 1995: 247, 255.
22 Gulliver 1955: 37–38; Spencer 1965: 5–6; 1988: 17–18; Dyson-Hudson 1966: 59, 112–133, 151; Legesse 1973: 86–87; Little 1985b: 139; Storas 1990: 139–141; Ensminger 1992: 131–132; Hogg 1993: 68.
23 Spencer 1973: 182–191; 1988: 18.
24 Cf. Swift 1977a: 173; 1977b: 464; Ensminger 1992: 130.
25 Gulliver 1955: 30–31, 39, 138, 260; Storas 1990: 138–139.
26 Spencer 1998: 252–267.
27 Lane 1998.

References

Anderson, D.M. 1989. Agriculture and irrigation technology at Lake Baringo in the nineteenth century. *Azania* 24: 84–97.

Baier, S. 1976. Economic history and development: Drought and the Sahelian economics of Niger. *African Economic History* 1: 1–16.

Baxter, P.T.W. 1993. The 'new' East African pastoralist: An overview. In *Conflict and the Decline of Pastoralism in the Horn of Africa.* (ed.) J. Markarkis. London: Macmillan. 143–162.

Baxter, P.T.W. and R.S. Hogg (eds). 1990. *Property, Poverty and People.* Manchester: University of Manchester.

Berntsen, J.L. 1979. Economic variations among Maa-speaking peoples. In *Ecological History.* (ed.) B.A. Ogot. Hadith 7. 108–127.

Bray, W. 1973. The biological basis of culture. In *The Explanation of Culture Change: Models of prehistory.* (ed.) C. Renfrew. London: Duckworth. 73–92.

Broch-Due, V. 1990. Livestock speak louder than sweet words: Changing property and gender relations among the Turkana. In *Property, Property and People.* (eds) P.T.W. Baxter and R.S. Hogg. Department of Anthropology, University of Manchester. 147–163.

Chambers, R.J.H. 1973. The Perkerra Irrigation Scheme: A contrasting case. In *Mwea: An irrigated rice settlement in Kenya.* (eds) R.J.H. Chambers and J. Moris. Munich: Weltforum Verlag. 344–364.

Dyson-Hudson, N. 1966. *Karimojong Politics*. Oxford: Clarendon.
Dyson-Hudson, N. and R. 1982. The structure of East African herds and the future of East African herders. *Development and Change* 13: 213–238
Ensminger, J. 1992. *Making a Market: The institutional transformation of an African society*. Cambridge: Cambridge University Press.
Galaty, J. 1980. The Maasai group ranch: Politics and development in an African pastoral society. In *When Nomads Settle*. (ed.) P.C. Salzman. New York: Praeger.
Gowlett, J.A.J. 1988. Human adaptation and long-term climatic change in northeast Africa: An archaeological perspective. In *The Ecology of Survival*. D.H. Johnson and D.M. Anderson. London: Lester Crook Academic Publishing. 27–45.
Gulliver, P.H. 1955. *The Family Herds*. London: Routledge and Kegan Paul.
—— 1963. *Social Control in an African Society: A study of the Arusha*. London: Routledge and Kegan Paul.
Hardin, C. 1968. The Tragedy of the Commons. *Science* 162: 1243–1248.
Hinde, S.L. and H. 1901. *The Last of the Masai*. London: Heinemann.
Hogg, R.S. 1993. Continuity and change among the Boran in Ethiopia. In *Conflict and the decline of pastoralism in the Horn of Africa*. (ed) J. Markarkis. London: Macmillan. 68–82.
Homewood, K.M. and W.A. Rodgers. 1987. Pastoralism, conservation and the overgrazing controversy. In *Conservation in Africa*. (eds) D.M. Anderson and R. Grove. Cambridge: Cambridge University Press. 111–128.
Houerou, H.N.Le. 1986. The desert and arid zones of northern Africa. In *Ecosystems of the World 12B: Hot deserts and arid shrublands*. (eds) M. Evernari, I. Noy-Meir and D.W. Goodall. Amsterdam: Elsevier. 101–141.
Jackson, F. 1930. *Early Days in East Africa*. London: Edward Arnold.
Jacobs, A.H. 1975. Maasai pastoralism in historical perspective. In *Pastoralism in Tropical Africa*. (ed.) T. Monod. London: Oxford University Press.
—— 1980. Pastoral Maasai and tropical rural development. In *Agricultural Development in Africa*. (eds) P.H. Bates and M.F. Lofchie. New York: Praeger. 275–300.
Johnston, H.H. 1886. *The Kilimanjaro Expedition*. London: Kegan Paul and Trench.
Kjekshus, H. 1977. *Ecology Control and Economic Development in East African History: The case of Tanganyika 1850–1950*. London: Heinemann.
Lamphear, J. 1976. *The Traditional History of the Jie of Uganda*. Oxford: Clarendon.
Lane, C.R. (ed.). 1998. *Custodians of the Commons*. London: Earthscan Publications.
Legesse, A. 1973. *Gada: Three approaches to the study of African society*. New York: The Free Press.
Little, P.D. 1985a. Social differentiation and pastoralist sedentarization in northern Kenya. *Africa* 55: 243–261.
——1985b. Absentee herd owners and part-time pastoralists: The political economy of resource use in northern Kenya. *Human Ecology* 13: 131–151.
—— 1992. *The Elusive Granary: Herder, farmer and the state in Northern Kenya*. Cambridge: Cambridge University Press.
Livingstone, I. 1977. Economic irrationality among pastoral peoples: Myth or reality? *Development and Change* 3: 209–230.
Markarkis, J. (ed.). 1993. *Conflict and the Decline of Pastoralism in the Horn of Africa*. London: Macmillan.
Mathias-Mundy, E. and C.M. McCorkle. 1995. Ethno-veterinary medicine and development: A review of the literature. In *The Cultural Dimension of Development:*

Indigenous knowledge systems. (eds) D.M. Warren, L.J. Slikkerveer and D. Brokensha. London: Intermediate Technology Publications. 488–498.

Merker, M. 1910. *Die Masai.* Berlin: Dietrich Reimer.

Niamir, M. 1995. Indigenous systems of natural resource management among pastoralists of arid and semi-arid Africa. In *The Cultural Dimension of Development: Indigenous knowledge systems.* (eds) D.M. Warren, L.J. Slikkerveer and D. Brokensha. London: Intermediate Technology Publications. 245–257.

Ndagala, D.K. 1990. Pastoral territoriality and land degradation in Tanzania. In *From Water to World-making.* (ed.) G. Pálsson. Uppsala: The Scandinavian Institute of African Studies.

Popper, K.R. 1963. *Conjectures and Refutations: The growth of scientific knowledge.* London: Routledge and Kegan Paul.

—— 1973. Evolutionary epistomology. In *A Pocket Popper*. (ed.) D. Miller. London: Fontana. 78–86.

—— 1977. Natural selection and its scientific status. In *A Pocket Popper*. (ed.) D. Miller. London: Fontana. 239–247.

Royal Commission on East Africa 1953–1955: Report. 1955. Cmnd 9745. London: HMSO.

Schlee, G. 1989. *Identities on the Move: Clanship and pastoralism in northern Kenya.* Manchester: Manchester University Press.

Sobania, N. 1980. *The Historical Traditions of the Peoples of the Lake Turkana Basin c.1840–1925.* PhD thesis, University of London.

Spencer, P. 1965. *The Samburu: A study of gerontocracy in a nomadic tribe.* London: Routledge and Kegan Paul.

—— 1973. *Nomads in Alliance: Symbiosis and growth among the Rendille and Samburu of Kenya.* London: Oxford University Press.

—— 1988. *The Maasai of Matapato: A study of rituals of rebellion.* Manchester: Manchester University Press.

—— 1998. *The Pastoral Continuum: The marginalization of tradition in East Africa.* Oxford: Clarendon Press.

—— 2003. *Time, Space and the Unknown: Maasai Configuration of power of providence.* London: Routledge.

Sperling, L. and J.G. Galaty. 1990. Cattle, culture and economy: Dynamics in East African pastoralism. In *The World of Pastoralism.* (eds) J.G. Galaty and D.L. Johnson. London: Guildford Press. 69–98.

Spooner, B. 1971. Towards a generative model of nomadism. *Anthropological Quarterly* 44: 198–210.

Storas, F. 1990. Intention of implication: The effects of Turkana social organization on ecological balances. In *Property, Property and People.* (eds) P.T.W. Baxter and R.S. Hogg. Manchester: Department of Anthropology, University of Manchester. 137–146.

Swift, J.J. 1977a. Desertification and man in the Sahel. In *Land Use and Development. African Environment Special Report 5.* (eds) P. O'Keefe and B. Wisner. London: International African Institute. 171–178.

—— 1977b. Sahelian pastoralists: Underdevelopment, desertification, and famine. *Annual Review of Anthropology* 6: 457–478.

Thomson, J. 1885. *Through Masai Land.* London: Samson Low.

Trimingham, J.S. 1968. *The Influence of Islam Upon Africa.* London: Longman.

Turton, D. 1975. The relation between oratory and the exercise of influence among the Mursi. In *Political language and Oratory in Traditional Society.* (ed.) M. Bloch. London: Academic Press. 163–183.

Tylor, E.B. 1865. *Researches into the Early History of Mankind and the Development of Civilization.* London: J. Murray.

Warren, D.M., L.J. Slikkerveer and D. Brokensha (eds). 1995. *The Cultural Dimension of Development: Indigenous knowledge systems.* London: Intermediate Technology Publications.

Index

For Product Safety Concerns and Information please contact our EU
representative GPSR@taylorandfrancis.com Taylor & Francis Verlag GmbH,
Kaufingerstraße 24, 80331 München, Germany

Printed and bound by CPI Group (UK) Ltd, Croydon, CR0 4YY
01/05/2025
01858587-0001